Geography SL&HL

FOR THE IB DIPLOMA

Urban Environments

Duncan McCulloch

PEAK

Published by:
Peak Study Resources Ltd
1 & 3 Kings Meadow
Oxford OX2 0DP
UK

www.peakib.com

Geography SL & HL Urban Environments: Study & Revision Guide for the IB Diploma

ISBN 978-1-913433-06-2

Books may be ordered directly from the publisher (see www.peakib.com) and through online or local booksellers. For enquiries regarding titles, availability or retailers, please email books@peakib.com or use the form at www.peakib.com/contact.

Printed and bound in the UK by:
CPI Group (UK) Ltd, Croydon CR0 4YY
www.cpibooks.co.uk

FSC www.fsc.org MIX Paper from responsible sources FSC® C013604

Cover image: aerial photo of an urban area with a mix of residential, recreational, commercial and public amenity land use (Adobe Stock)

Preface

The year 2007 marked the culmination of one of the greatest shifts in demographic history: for the first time more people lived in urban areas than rural areas. With the world population set to reach 9.8 billion by 2050 and 68% of those predicted to reside in urban areas, this topic will become even more relevant. Add in the complicated dynamics of the faster growth of population in Low Income Countries (LICs) and their high rates of rural-urban migration then the necessity to create sustainable urban living environments is clear.

This study and revision guide has been written to help you develop your understanding of the urban environments option, which has direct links to the core themes of the IB Geography course: population, climate change, and resource consumption. There are supporting case studies and examples included throughout and you should attempt all of the questions in preparation for your exam. Suggested answers are provided at the back of the book for you to check afterwards and there are boxes in the margin containing exam tips, resources for extra research (time allowing!) and advice. Where some words or key terms may not be familiar to you, I have highlighted them in **bold** in the text and incorporated definition boxes on the relevant pages.

Your IB Geography Paper 1 exam will contain a structured question on urban environments as well as a choice of two essays. I hope this guide will help bring you success in this exam, but furthermore I hope studying this option has inspired you to consider some of the crucial aspects of modern urban society and how you may adapt your living in the future. I have visited and taught in various cities worldwide and have tried to use personal experiences and case studies where applicable although, given the global nature of this topic and demands of the syllabus, this has not always been possible. You should also be aware that using personal examples is encouraged in this option and the wider geography syllabus and you may be able to use the city you live in for many of the examples.

My thanks go to Ben Thatcher, Garett Nagle, Stuart McCulloch, and Nick Bell and the Peak team for making this publication possible.

Duncan McCulloch

Contents

Introduction

For Option G: Urban Environments, the *Geography Syllabus Guide 2019* identifies specialist concepts for analysis including:

- Hierarchies;
- Systems;
- Sustainability.

The guide considers urban places as 'sites of intense social interaction and as focal points of production, wealth generation and consumption.' In addition to this, the contrasts of wealth and deprivation and the necessity of minimising harmful social and environmental stresses play a key role in the current thoughts of urban planners.

The syllabus breaks the option down into four thematic sections:

- The variety of urban environments;
- Changing urban systems;
- Urban environmental and social stress;
- Building sustainable urban systems for the future.

You are required to have specific case studies and should use examples that you have studied in class or researched yourself. I have used my own examples to explore the themes within this guide:

CASE STUDY THEME	EXAMPLES REQUIRED	EXAMPLE IN THIS BOOK
The infrastructure growth over time in a city	One	Rio de Janeiro, Brazil
Air pollution and its varying impact on people	One	Delhi, India
Traffic congestion in a city and the management response	One	Brussels, Belgium
Land use conflicts in contrasting neighbourhoods	Two	Kumasi, Ghana Barra de Tijuca, Brazil
Resilient city strategies	Two	Melbourne, Australia Cities in the European Union (EU)
Environmental strategies to create eco-cities	Two	Växjö, Sweden Vancouver, Canada

Furthermore, general place examples are expected in your work and there are several featured in this book including:

> Barcelona, Spain; Belmont, USA; Chicago, USA; Dhaka, Bangladesh; Glasgow, Scotland; London, England; Lusaka, Zambia; Toronto, Canada; unnamed new planned capital city, Egypt.

About this book

This book is not intended as a substitute for your textbook, class work or independent research; it is more of an additional, focused guide that takes you through the key parts of each topic, provides advice and pointers to other resources or linked parts of the syllabus, and is especially useful for exam preparation.

It is divided into four main chapters mirroring the structure of the syllabus, with an emphasis on revision questions and answers rather than detailed content. Each chapter begins with a reminder of the key question you are expected to address and a list of the geographic knowledge and understanding points laid out in the syllabus.

The examples and case studies that are used throughout the book should be further developed with your own research.

Navigating the guide

This guide is structured to enable you to study efficiently:

- essential information is highlighted in separate boxes,
- cross references provide links between sections or external resources, and
- practice questions follow the relevant sections of text.

We use icons to help you quickly and easily identify different types of information.

Key to icons used in this study guide

Author's study note or exam tip

Case study/worked example

Practice question

Key term

Key information

More information

Cross-reference

Online media link

Underlying concepts

The IB Geography course has been designed around a '4 Ps' concept of Place, Processes, Power and Possibilities. These should be considered when applying your knowledge, along with the Sustainable Development Goals (SDGs).

Almost all these goals, launched in 2015 and to be achieved by 2030, link into this topic; particularly relevant are numbers 8, 9, and 11.

Figure i: The Sustainable Development Goals

Source: United Nations. 2016. www.un.org/sustainabledevelopment (accessed 9 Sept 2021)

Exam guidance

The Paper 1 exam is 2 hours 15 minutes (HL) or 1 hour 30 minutes (SL). This means you should spend around 45 minutes on this option, but you will need to time yourself as no reminders will be given.

You will have 20 marks worth of questions for Urban Environments, out of a total of 60 (HL) or 40 (SL). This paper is worth 35% of your total IB Geography score. Basically, every mark counts!

You will have 10 marks of shorter answer questions. Expect some of these to be skills based, either describing graphs (using quantification), or possibly map interpretation (as in the Mapping Exercise at the end of this guide). The others will be based around any of the content in the syllabus. Ensure you know what each command term is asking you (is it 'explain', 'analyse', 'outline', 'state', etc?) and that you don't waste time writing too much in these answers. If it asks for 2 marks, generally expect to make two valid points. Questions can also be structured as 2+2 so be aware you should expand on your answers/give examples as relevant.

The other 10 marks are to be written in essay format. Ensure you are comfortable with the IB mark bands and that you take into account essay structure as well as content. It is advisable to make a short plan, but be aware you should be writing this essay in under 25 minutes so don't spend too long on this. Command terms are often 'examine' or 'evaluate'. Aim for around five paragraphs with an introduction, main body paragraphs, and a short conclusion. Your paragraphs would ideally have different themes and bring in case study detail. You should always look at evaluating within your response.

Practising timed essays, hand written (unless you type in the exam), is highly recommended. Many of the practice questions you will answer in this guide could be written in essay format: for example, Section 1.3 practice Q6; Section 2.6 practice Q4; Section 3.3 practice Q2; and Section 4.8 practice Q2.

Abbreviations

BRT	Bus Rapid Transit
CBD	Central Business District
EU	European Union
GDP	Gross Domestic Product
HIC	High Income Country
HMIC	Higher Middle Income Country
LIC	Low Income Country
LMIC	Lower Middle Income Country
PLVI	Peak Land Value Intersection
PM	Particulate Matter

Chapter 1: The Variety of Urban Environments

Overview

This chapter outlines the characteristics of urban places, looking at their site, function, land use and hierarchy. Furthermore, the growth of urban places into megacities is examined, along with patterns of urban economic activities and residential areas. Finally, the incidence of poverty in cities at different levels of economic development is investigated.

Ensure when studying this topic online your resources are following the syllabus for exams beginning 2019 and not 2011. There is a lot online and this is an easy mistake to make!

KEY QUESTION

How do the characteristics and distribution of urban places, populations and economic activities vary around the world?

Geographic knowledge and understanding points

- Early and future development of urban places including land use and function
- The hierarchy of settlements and rise of the megacity
- The contrast of poverty incidence between cities at varying levels of economic development

1.1 Characteristics of Urban Places: Site and Function

When describing or explaining the evolution of urban areas it is important to consider both the necessary features of human life and what would be needed to 'create' an urban place. This is known as the **site** and **situation** of a settlement, with the site being the original physical (natural) factors enabling a settlement to grow. The situation of a settlement is then important in determining whether it grows to become a large city or stays as a small town or village, as it includes aspects such as other settlements, infrastructure and features around it. Figure 1.1 shows potential sites for early urban development. Think about why each site might be more or less suitable. Some of the key factors for choosing where to place a settlement are:

- water supply, but not at the risk of flooding
- defence from potential enemies
- near to a coastline for trading opportunities
- flat land to build upon
- fertile soils for farming
- supply of natural resources
- favourable climate and aspect (south facing)
- local access to natural resources (within walking distance)
- access to other forms of infrastructure such as bridges.

Section 1.1 practice Q1

Figure 1.1: **Sites of potential early settlements**

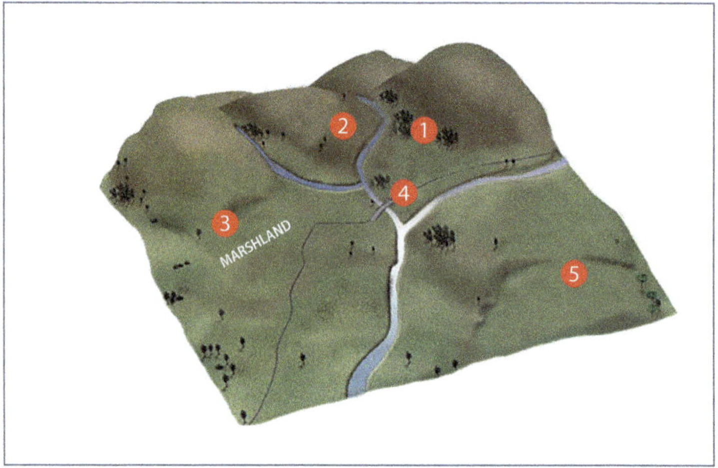

It is no coincidence that a majority of modern urban environments are located near a river or a coastline. Figures 1.4 and 1.5 show world maps of current major urban places.

With the rise of technology in the twenty-first century meaning work and services can often be brought to the home, the concept of **distance decay** largely becomes irrelevant in HICs. Therefore, many of the original site and situational factors are defunct today and city planners are faced with different issues. Factors such as sustainability and the location

It is easy to use throwaway statements without considering the differences between cities in LICs and HICs. Distance decay can certainly still be a factor in LICs.

Site: The physical land on which a settlement is built.

Situation: The relationship with surrounding human features.

Distance decay: The concept that as distance between places increases, interaction declines.

and development of new 'smart' or 'eco' cities have become increasingly important (see sections 4.6–4.8 for further detail on the development of new 'smart' and 'eco' cities).

The **function** of settlements is their primary role. As cities began to develop in the mid-1700s specific functions emerged, such as industry, market, port or resort. Today's larger urban places rarely have one specific function, but a combination of many.

Section 1.1 practice Q2

FUNCTION	MEANING
Industry	An urban area that developed with industry being their primary activity.
Market	An urban area that developed with selling and buying goods at a market being their primary activity.
Port	An urban area that developed due to trading.
Resort	An urban area that developed due to tourism.

Table 1.1: **Functions of settlements**

SECTION 1.1 PRACTICE

1. Study Figure 1.1 on page 2. Describe the advantages and disadvantages of sites 3 and 4.

2. Look again at Table 1.1. Two examples are shown below for Industry; add some more of your own for each type.

FUNCTION	EXAMPLES
Industry	Glasgow, UK Chicago, USA
Market	
Port	
Resort	

Function: The primary role of a settlement.

1.2 Characteristics of Urban Places: Land Use and Hierarchy

Land use can vary considerably between different urban areas, but uses have been portrayed through various models. These include Burgess 'Concentric Rings,' Hoyt 'Sectors' (figures 1.3 and 1.4) and the Harris and Ullman 'multiple nuclei' model (see Section 1.2 practice Q4).

Although somewhat simplistic and generalised these still have some use, particularly when understanding **bid-rent theory** and a city's **core-periphery pattern** (these concepts are further developed in sections 1.4 and 1.5).

If you search online you will find many examples of urban models. Some have validity, while others have little relevance to more than one city so be careful if you are using them.

Figure 1.2: **Concentric Land Use Model**

Source: Based on Burgess, Ernest W. 1925/67. 'The Growth of the City.' In Robert E. Park, Ernest W. Burgess, and Roderick D. McKenzie, *The City*. Chicago: University of Chicago Press. pp 47-62

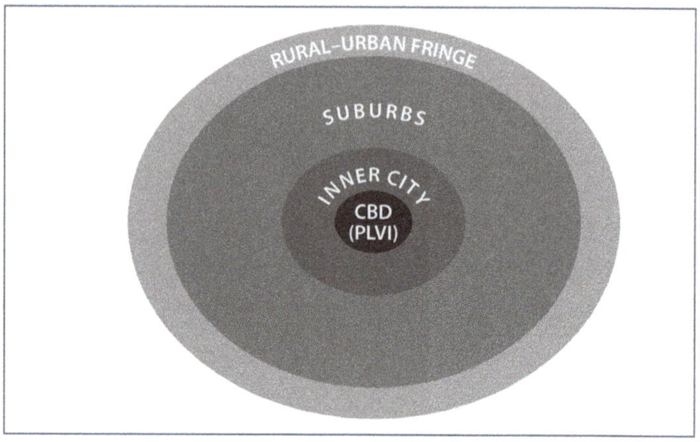

Figure 1.3: **Sector Land Use Model**

Source: Based on Hoyt, Homer. 1939. *The Structure and Growth of Residential Neighbourhoods in American Cities*. Washington, DC: Federal Housing Administration

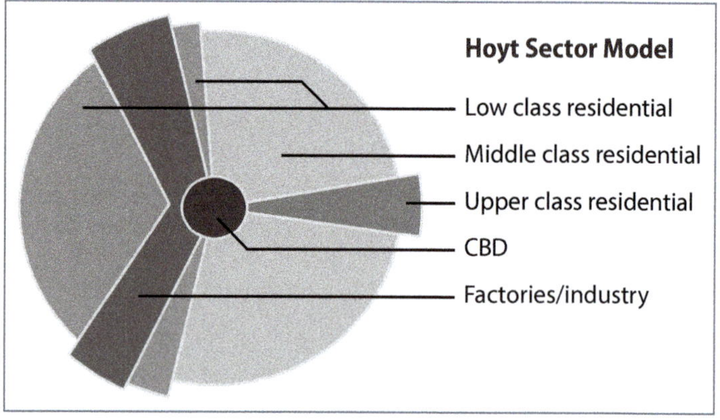

A land use acronym is **RICEPOTS**. What could that stand for? See Section 1.2 practice Q5.

Acronyms can be an excellent way to remember key points of information for your exam. For example: **SEEP**, when constructing essays, considering the **S**ocial, **E**conomic, **E**nvironmental and **P**olitical factors.

Bid-rent theory: The theory that the highest bidders can pay the highest rent. In urban places this indicates the central, most accessible area will have the highest rent.

Core-periphery pattern: The world is said to be divided into core regions (mainly HICs) and periphery regions (mainly LICs). Cities can also have core areas (usually the CBD) and peripheral areas (outskirts).

CBD: Central Business District. The town centre or downtown area, with a variety of services.

PLVI: Peak Land Value Intersection. The highest land value in the city.

Rural-urban fringe: The 'edge of the city,' often breaking into green, rural areas.

With the variety of functions and land uses, a **hierarchy** showing the size of settlements has developed. Whilst there are various ways of demonstrating this, Figure 1.4 shows a pyramid example, with a **megacity** being placed at the top, and an isolated dwelling at the bottom. Other terms such as **metropolitan area** or **megalopolis** or the proposed 'super urban regions' in China could also be added to this as urban areas continue to grow. The model has been partly annotated to discuss potential economic activities.

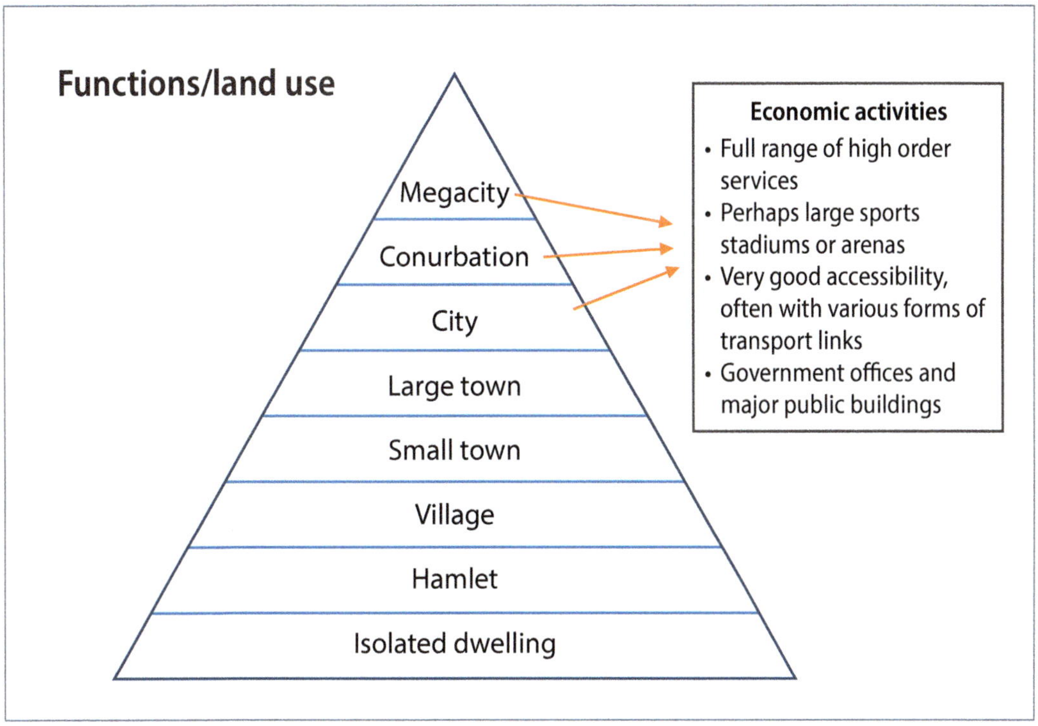

Figure 1.4: **Hierarchy of settlements**

The larger the settlement and its **threshold** population, the **higher order of services** and larger **sphere of influence** it is likely to have.

Hierarchy: Placing items in order of importance/size.

Megacity: A city with over 10 million people.

Metropolitan area: A region consisting of a densely populated urban core and its less-populated surrounding territories, including the rural-urban fringe.

Megalopolis: A chain of adjacent metropolitan areas, which may merge into a continuous urban region.

Threshold: A threshold population is the minimum number of people necessary before a particular good or service can be provided in an area.

High order services: Services that are bought less often, e.g. cars.

Sphere of influence: A spatial region over which an urban place can draw its customers/visitors.

SECTION 1.2 PRACTICE

1. Suggest possible functions and land uses for:

   (a) A Megacity

   ....

   ....

   ....

   ....

   (b) A large town

   ....

   ....

   ....

   ....

   (c) A small town

   ....

   ....

   ....

   ....

2. Explain how differing settlements can be placed into an ordered 'hierarchy.'

   ....

   ....

   ....

   ....

   ....

3. Analyse the impact of site and situation on the land use and functions of a city.

   ....

   ....

   ....

   ....

   ....

4. Figure 1.5 illustrates a multiple nuclei model.

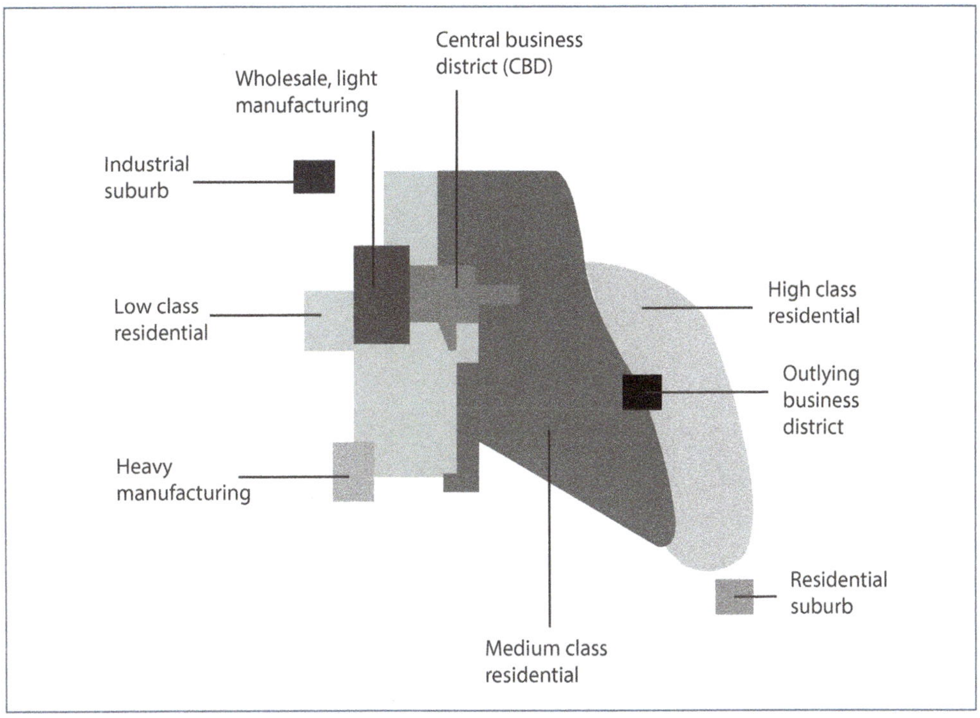

Figure 1.5: **Harris and Ullman's Multiple Nuclei Model**

Source: Based on Harris, C. D., & Ullman, E. L. 1945. 'The Nature of Cities,' in *The ANNALS of the American Academy of Political and Social Science*, 242(1), 7–17

Explain its characteristics and its relevance for cities today.

....

....

....

....

5. Land can be used in a number of ways in an urban area. What could the acronym RICEPOTS stand for? The first one has been done for you.

R esidential

I

C

E

P

O

T

S

1.3 Megacities and the Urban Growth Process

> Andrew Marr's documentary series 'Megacities'(2009) is excellent viewing, as is the more recent series 'Megacities of the World' by Bryce Plank (2019).

The 'rise of the megacity' (cities with over 10 million people) in the twenty-first century has been well documented and by 2019 there were a reported 37 of them on earth, with Tokyo, Jakarta, and Delhi making up the top three (see Figure 1.6). These tend to be dynamic urban places with considerable successes and stresses.

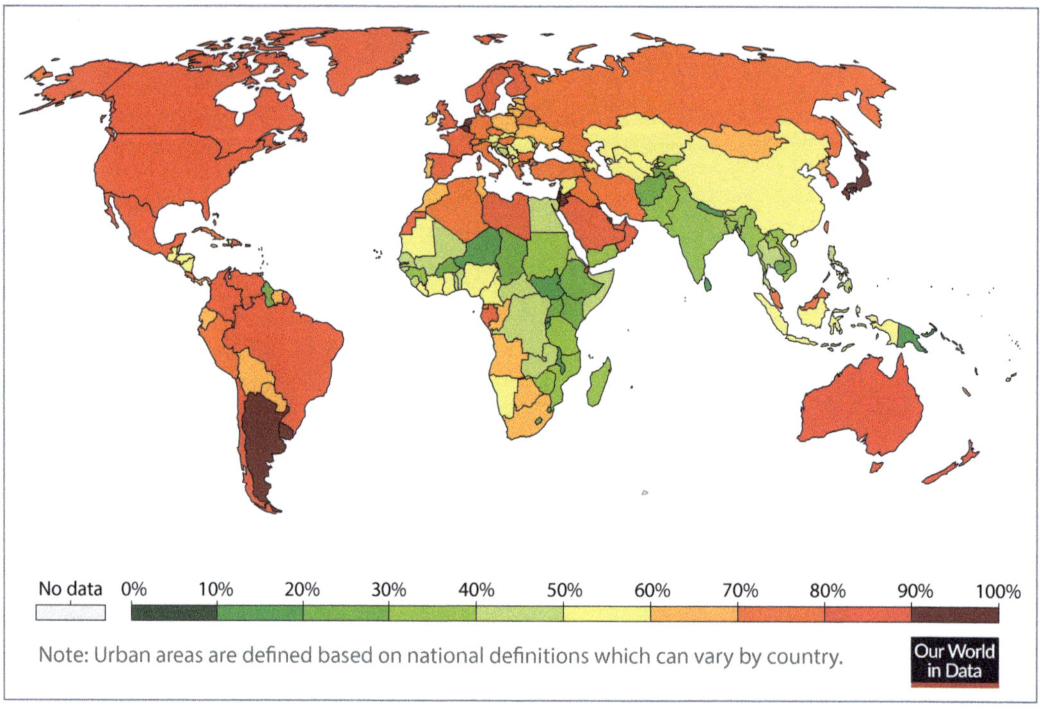

Figure 1.6: **Share of total population living in urban areas, 2017**

Source: Hannah Ritchie and Max Roser. 2019. 'Urbanization'. https://ourworldindata.org/urbanization (accessed 04/03/2019). CC-BY

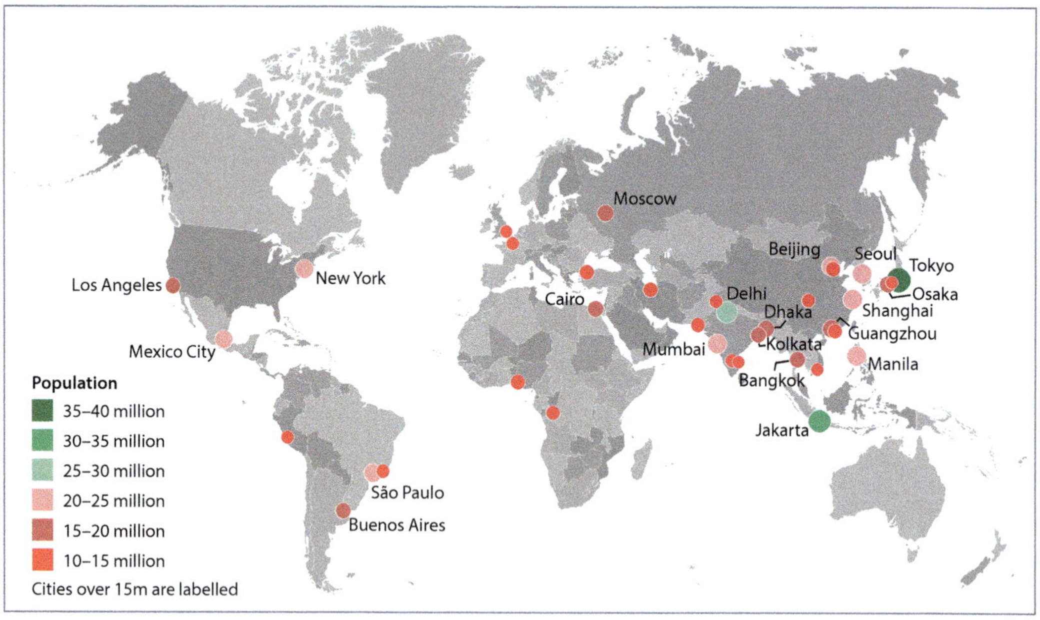

Figure 1.7: **The location of the 37 megacities, 2018**

Source: Data from: http://demographia.com/db-worldua.pdf (accessed 03/04/2019)

The urban growth process can be either **planned** or **spontaneous**.

Planned cities are those that are carefully planned from their inception, often on new, undeveloped land. Brasilia, Brazil (as a new capital city) and Chandigarh, India (as an architectural marvel and capital of two regions) are examples of this.

More recently, Egypt is replacing Cairo with a new planned administrative capital, similar to Brasilia, which will house five million residents (compared to 20 million in the Cairo region). Although not yet named, it is mooted as the 'biggest planned city ever' and promises a greener, cleaner and more welcoming environment.

When it is fully functional it plans to have:

- a new parliament
- a central bank
- an airport
- a presidential palace (eight times bigger than the White House)
- a business district
- Africa's tallest tower
- Egypt's tallest minaret and church steeple
- a theme park bigger than Disneyland.

With large urban growth projections for 2050, these planned cities are becoming more common (see section 4.1 for further detail on large urban growth projects for 2050). Although not officially a modern 'planned city', Figure 1.8 shows the skyline of Frankfurt, Germany, redeveloped and carefully planned as a financial centre following destruction in World War II.

You can read more about the plans for the new Egyptian capital city at https://constructionreviewonline.com/project-timelines/egypts-new-administrative-capital-project-timeline-and-what-you-need-to-know/ and at https://www.aljazeera.com/amp/opinions/2021/7/5/why-is-egypt-building-a-new-capital

Another ambitious example is in Equatorial Guinea where the government moved their headquarters to the Ciudad de la Paz ('City of Peace') in the rainforest in 2017: https://www.bbc.com/news/world-africa-38911573

Figure 1.8: **Skyline of Frankfurt, Germany**

Spontaneous urban growth is fast and unplanned and is particularly evident in cities that are rapidly growing in LICs and MICs such as Lagos, Dhaka (see sections 2.1 and 4.2) and Bangkok (see Figure 1.9). Newly arrived migrants from rural areas often struggle to afford the high cost of urban living and set up home in shanty towns or slums that spontaneously expand the city limits (see Figure 1.10).

HIC cities have also seen substantial spontaneous growth and **urban sprawl** is a common issue, seen for example in Paris, France. As the city grew rapidly during the twenty-first century, processes such as suburbanisation led to the city struggling to maintain its city limits (the concept of suburbanisation is explored further in section 2.3).

The website https://www.citylab.com/equity/2014/04/visualizing-200-years-urban-expansion-paris-sao-paulo-la/8794/ has some interesting map animations showing urban sprawl over time.

Figure 1.9: **Skyline of Bangkok, Thailand**

Figure 1.10: **Spontaneous urban settlement appearing at the edge of Johannesburg, South Africa**

SECTION 1.3 PRACTICE

1. Study Figure 1.6 on page 8. State the continent with:

   The highest % share of urban population

   ....

   The lowest % share of urban population

   ....

2. Study Figure 1.6. Identify three countries with an urban population below 60%.

   ....

   ....

   ....

3. Suggest two possible reasons for one of the countries you identified in question 2.

....

....

....

....

4. Study Figure 1.7. Describe the global distribution of megacities in 2018.

....

....

....

....

5. Study Figure 1.8 and Figure 1.9. Compare and contrast the characteristics of the two skylines.

For questions with the command term 'compare and contrast', you are expected to provide both similarities and differences.

....

....

....

....

6. Referring to a specific planned city, outline the benefits of the new settlement.

....

....

....

....

7. Analyse the factors leading to the growth of spontaneous settlements.

....

....

....

....

1.4 Factors Affecting the Pattern of Urban Economic Activities

Urban economic activities can refer to retail (shopping), commercial (business), or industrial. These can be looked at in conjunction with the urban land use models (refer back to section 1.2 for further information on urban land use models).

LAND USE ZONE	MAIN ECONOMIC ACTIVITIES	CHARACTERISTICS
CBD	Retail, commercial	High and low order goods, high land value, densely built
SUBURBS	Residential, retail	Low order goods, residential
RURAL–URBAN FRINGE	Residential, retail, industrial	High and low order goods, lower land value, more open space

Table 1.2: **Pattern of urban economic activities**

The bid-rent theory (Figure 1.11) indicates that land value is at its highest in the most accessible, central part of the city, and therefore the highest order goods and variety of retail is found in the CBD. This also leads to **vertical zoning**. However, as land value is cheaper in the suburbs and edge of city locations, it has been common for large retail companies (such as IKEA) to set up their main shops there. Ease of parking and lower pollution rates are other factors; although, as the planet becomes more urbanised, these factors are also said to be a catalyst for urban sprawl and more spontaneous development.

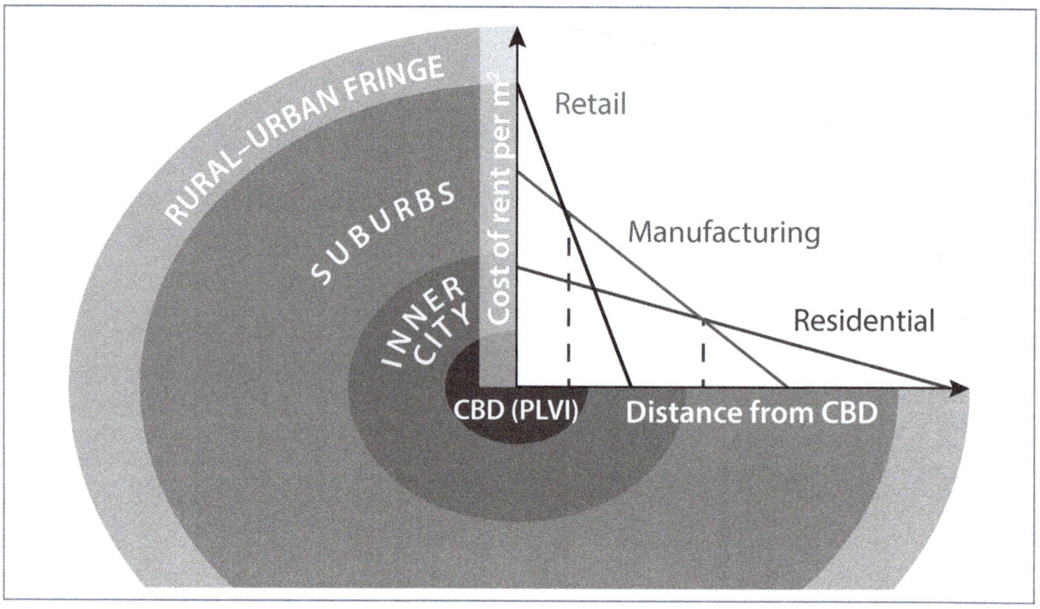

Figure 1.11: **Bid-rent theory**

Physical factors such as being located near a river and natural resources also play a role in the pattern of economic activities. This is clearly evident in older settlements such as London and New York with historic economic activities being focused around the use of water for trade.

Vertical zoning: Due to the high price of land, developers tend to build upwards in the CBD, perhaps with different land uses in the same skyscraper, such as retail and commerce.

1.5 Factors Affecting the Pattern of Residential Areas within Urban Areas

The pattern of residential areas varies according to both the age and economic development of the associated urban area. Similar to economic activities, physical factors such as water and relief play their role, often acting as barriers to building. However, many spontaneous settlements, such as the **favelas** in Rio de Janeiro or the townships in Johannesburg (Figure 1.10 on page 10), appear on steep mountain sides as the only available free land people can build on.

Favela is the Brazilian term for a shanty town.

Land value is also a considerable factor, with higher cost residential zones in HICs springing up in both suburban areas and gentrified inner city areas (the gentrification of HICs is covered in further detail in section 2.2). Lower cost housing is often congregated around the CBD and in inner city locations, although residential density increases the further from the centre as illustrated in the bid-rent theory (Figure 1.11). In LICs they often follow the more 'sector' pattern, as laid out by the Hoyt model (Figure 1.3 on page 4), following transport or industrial lines and particularly where there is an abundance of shanty housing.

An entertaining travel blog written in late 2018 giving an account of contemporary life in a favela from a Western point of view is at: https://www.travelsauro.com/living-favela-brazil/

Ethnicity can play a role in determining residential areas, and it is common to find 'ethnic clusters' in cities. Examples of this could be the Congolese district of Matonge, Brussels; the Indian community in Harrow, London; or the large Chinatowns seen throughout cities in North America. Furthermore, wealthy 'gated' communities are still a common site in post-apartheid South African cities.

Having studied planned settlements in section 1.3, it is evident that specific urban residential planning also has an impact on the pattern of residential areas, as illustrated in Figure 1.8 on page 9.

Linking units of work together can help in your understanding and answering of exam questions.

SECTIONS 1.4 AND 1.5 PRACTICE

1. Study the chart below. Describe and explain the differences in land use seen from the CBD to rural-urban fringe in this HIC city.

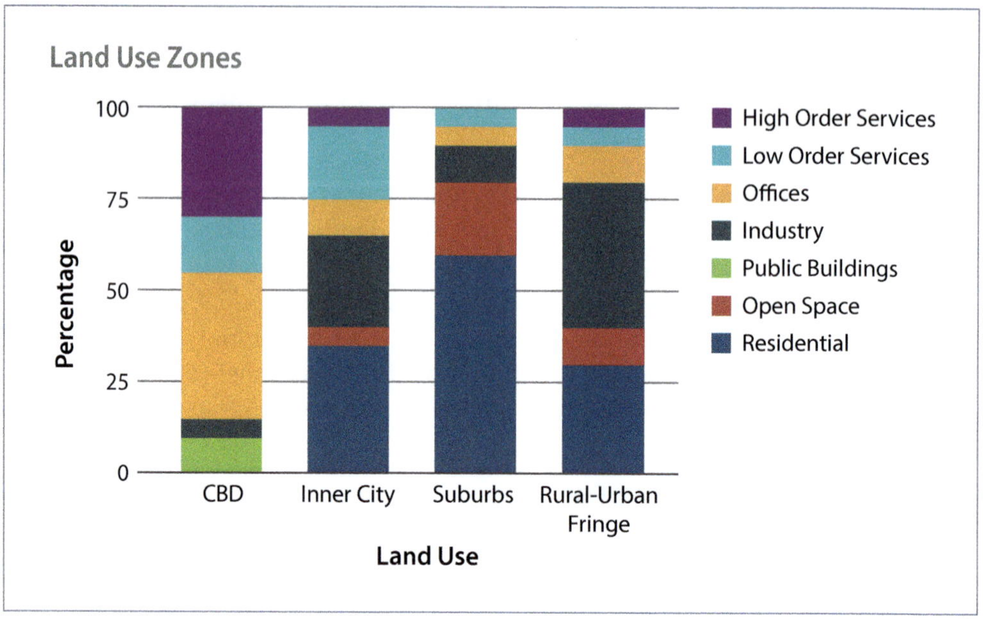

2. Outline the reasons why 'ethnic residential clusters' can form in urban areas.

3. Explain the 'bid-rent theory'.

4. Discuss the relative importance of physical and human factors in determining the pattern of:

   (a) Economic activities

   ....

   ....

   ....

   (b) Residential areas

   ....

   ....

   ....

1.6 The Incidence of Poverty, Deprivation and Informal Activity in LICs

Poverty, according to the UN (2018), is 'more than the lack of income and resources to ensure a sustainable livelihood. Its manifestations include hunger and malnutrition, limited access to education and other basic services, social discrimination and exclusion as well as the lack of participation in decision-making'. **Absolute poverty** is when household income does not meet the minimum requirements for basic living needs, generally calculated as less than $1.90 per day. To this end, 'no poverty' is the first Sustainable Development Goal to be achieved by 2030.

Deprivation is the lack or denial of something deemed necessary and is therefore directly linked to poverty incidence. **Informal activity** is 'work involving the paid production and sale of goods or services which are unregistered by, or hidden from the state, for tax and/or benefit purposes, but which are legal in all other respects' (Williams, 2004).

Almost every urban area in the world has an element of poverty, but at different scales. It is evident that as populations become more urban the **GDP** of the country is generally increased, and therefore overall poverty levels decline, as illustrated in Figure 1.12.

Absolute poverty: Where household income is below a necessary level to maintain basic living standards, generally accepted as less than $1.90 per day.

Deprivation: Being without a basic, necessary need.

Informal activity: Work done that is neither taxed nor monitored by the government.

GDP: Gross Domestic Product. The total value of goods and services produced within a country.

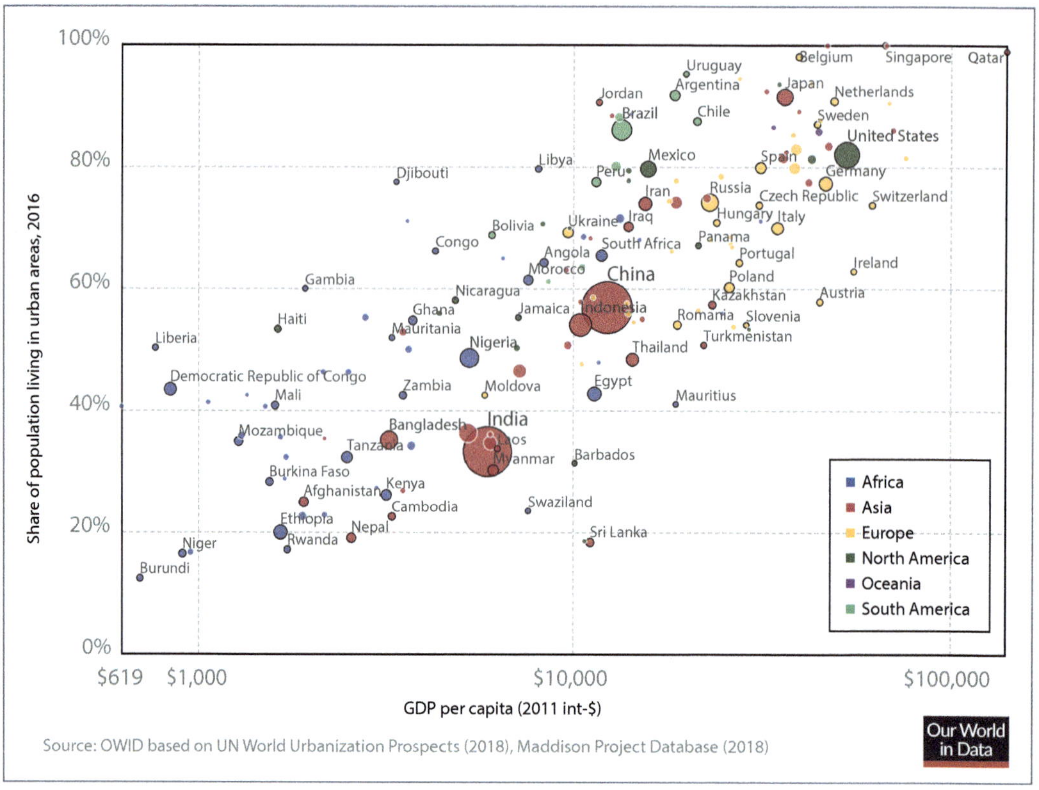

Figure 1.12: **Levels of urbanisation and GDP**

Source: Hannah Ritchie and Max Roser. 2019. 'Urbanization'. https://ourworldindata.org/urbanization (accessed 04/03/2019). CC-BY

CASE STUDY

Lusaka

Lusaka is the capital of Zambia, located south of the Equator in Sub-Saharan Africa. Between 1990 and 2010 its population more than doubled from 757,000 to 1.7 million. 70% of these inhabitants live in slums without proper access to basic services such as water and sanitation (De la Fuente *et al*, 2015). Disease is common: one of the largest slums, Kanyama, declared a curfew in late 2017 following a cholera outbreak linked to contaminated food and water.

Figure 1.13: **Poverty incidence in Zambia**

Source: De La Fuente et al. 2015. https://openknowledge.worldbank.org/handle/10986/21783 (accessed 04/03/2019). p. 32. © World Bank. CC-BY-3.0 IGO

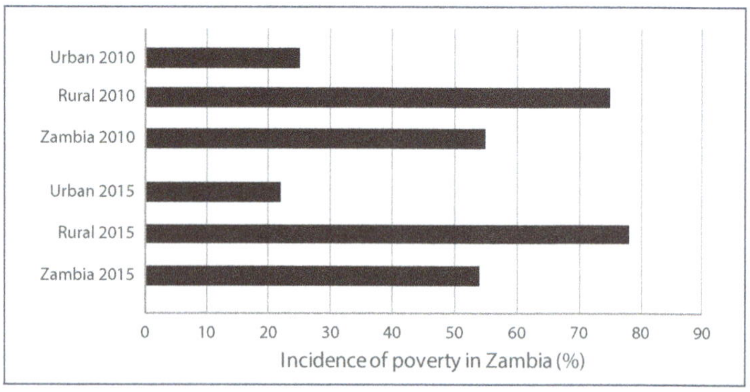

The slums are in plentiful supply due to the high rates of rural-urban migration typical of LICs (see section 4.2 on trends of urban-rural migration). Figure 1.13 shows the incidence of poverty in Zambia in 2010 and 2015 and illustrates why people are still migrating towards urban areas. However, Luangwa district in Lusaka is widely regarded to be in the worst five areas for deprivation in Zambia (as reported in the South African Regional Poverty Network) and Figure 1.14 also shows that despite overall lower rates of poverty in urban areas, Lusaka has some of the highest poverty rates in the country.

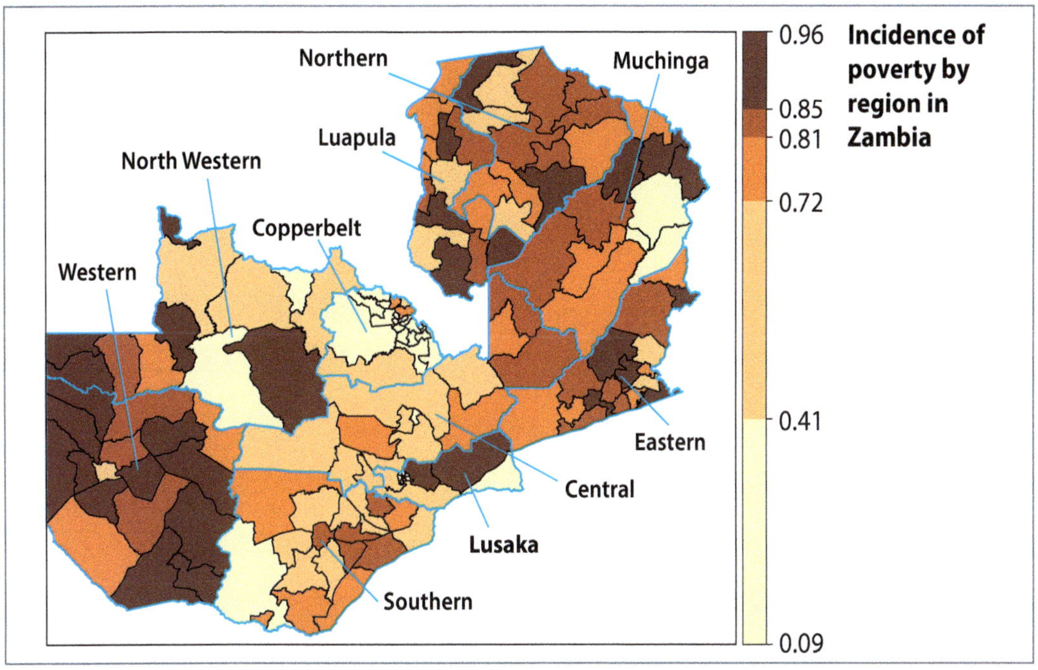

Figure 1.14: **Incidence of poverty by region in Zambia**

Source: Data from De La Fuente *et al*, 2015. https://openknowledge.worldbank.org/handle/10986/21783 (accessed 04/03/2019)

In Lusaka, 61.6% of all jobs are based around the informal sector. This ranges from construction, trade, transport and 'other' service activities (Kalumbi, 2016). With the lack of tax revenue generated from this type of employment it continues to be a huge effort for the government to raise GDP and overall living standards.

It is often assumed that all work in the informal sector is illegal; however, this is not true and often depends on the country.

SECTION 1.6 PRACTICE

1. Describe the incidence of poverty in Zambia.

2. Explain why the informal economy plays such a large role in lower income cities such as Lusaka.

....

....

....

....

....

3. Suggest reasons why poverty tends to be reduced with increased rates of urbanisation.

....

....

....

....

....

....

....

1.7 The Incidence of Poverty, Deprivation, and Informal Activity in HICs

CASE STUDY

London

In November 2018, the UN poverty envoy estimated that a fifth of the UK population were living in poverty. The majority of this poverty is centred in urban areas and particularly parts of London, where approximately 27% of the population live in **relative poverty** (as opposed to absolute poverty). This is measured by being in a household with an income below 60% of the median. Although this is adjusted for household size, the high cost of rent and housing in the city means this number is well above the UK average.

According to the London Poverty Profile of 2017, the top 10% of households received nearly 30% of income but owned just over 50% of total wealth. The bottom 50% of

You can read more on this here: https://www.trustforlondon.org.uk/data/

Relative poverty: The condition in which people lack the minimum amount of income needed in order to maintain the average standard of living in the society in which they live.

Londoners received nearly 25% of income but owned only 5% of wealth. This wealth gap creates clear districts or 'boroughs' that have a higher incidence of deprivation and poverty than others (see Figure 1.15 and Figure 1.16).

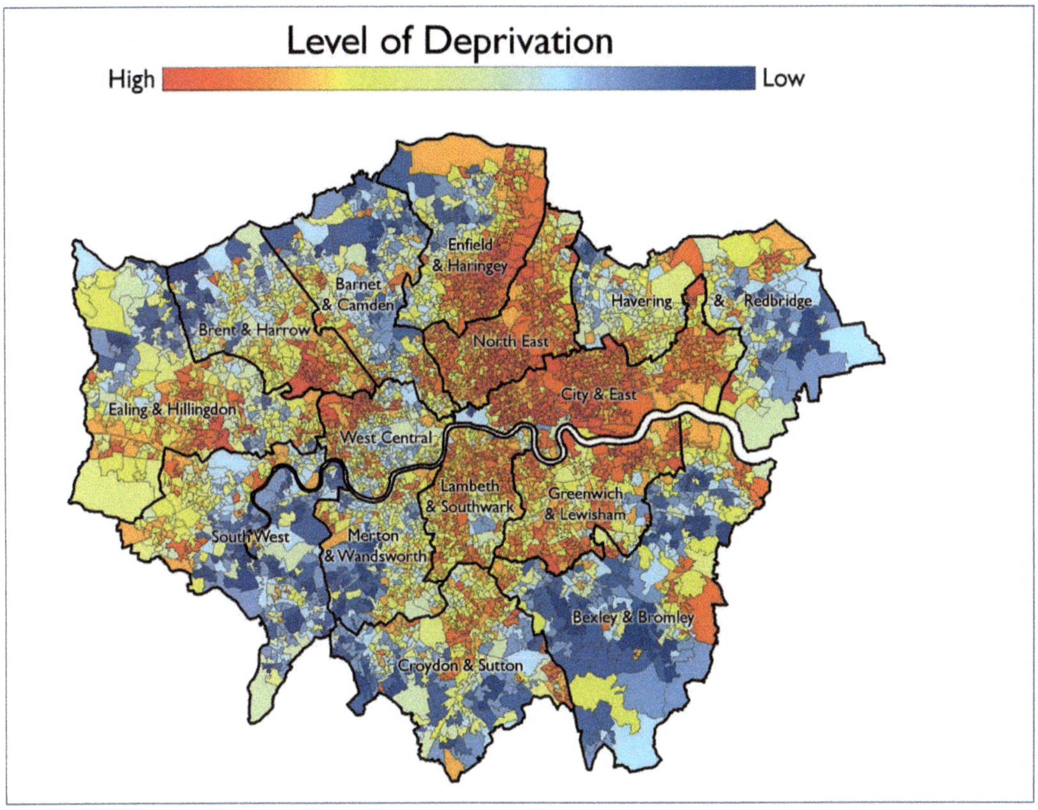

Figure 1.15: **Levels of deprivation in London**

Source: Alasdair Rae. 2012. www.statsmapsnpix.com. Used with permission of Alasdair Rae

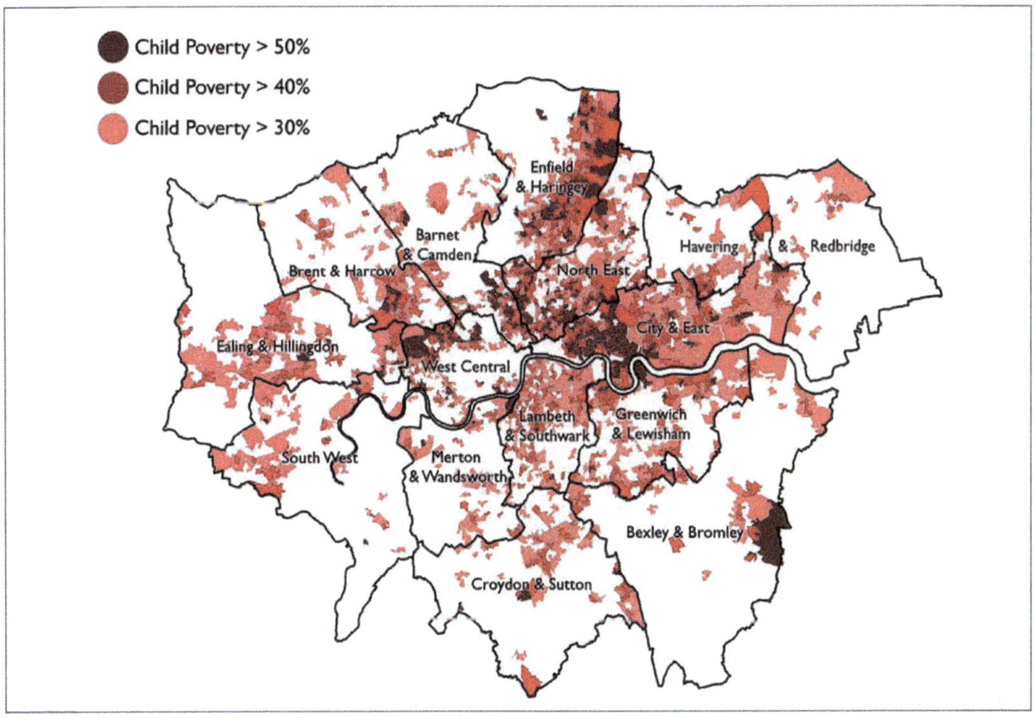

Figure 1.16: **Levels of child poverty in London**

Source: Alasdair Rae. 2012. www.statsmapsnpix.com. Used with permission of Alasdair Rae.

The informal sector, although far smaller than that in Lusaka, still plays a prominent role in the UK, with around 10% of national income thought to come from this. The numbers are similar in London and come from mainly the construction, caring and gardening sectors (Elliot, 2013).

SECTION 1.7 PRACTICE

1. Using examples, compare and contrast the incidence of poverty in cities from LICs and HICs.

2. Study Figure 1.15 and Figure 1.16. Describe the patterns of child poverty and deprivation in London.

3. Explain why it can be difficult to define 'poverty'.

Chapter 2: Changing Urban Systems

Overview

Urbanisation is the increase in the percentage of people living in urban areas. This can stem from natural population increase (when the birth rate is higher than the death rate), from **centripetal** population movements or from reclassification of formerly rural areas.

Urban population movements can be considered inward or outward. While those that move into the city are considered centripetal, those that move outwards are known as **centrifugal**.

Students frequently get these two movements mixed up – can you think of a way to remember them?

KEY QUESTION

How do economic and demographic processes bring change to an urban area over time?

Geographic knowledge and understanding points

- Inward and outward urban movements
- Urban system and infrastructure growth
- Deindustrialisation causes and consequences

Centripetal: An inward urban movement.

Centrifugal: An outward urban movement.

2.1 Centripetal Population Movements in Industrialising Cities

Centripetal movements in industrialising cities are characterised by rural-urban migration as people search for a higher quality of living and better paid jobs.

Despite Africa being the most rural continent at a reported 57% in 2018 (UN, 2018), rapidly industrialising cities (such as Lusaka) are currently urbanising the fastest. Figure 2.1 shows the impact of **rural-urban migration** and natural increase on the population of various world cities. This migration is characterised by **push factors**, such as uneconomic farming in rural areas, and **pull factors**, most notably job opportunities in urban areas. More on the trends of this migration are in section 4.2.

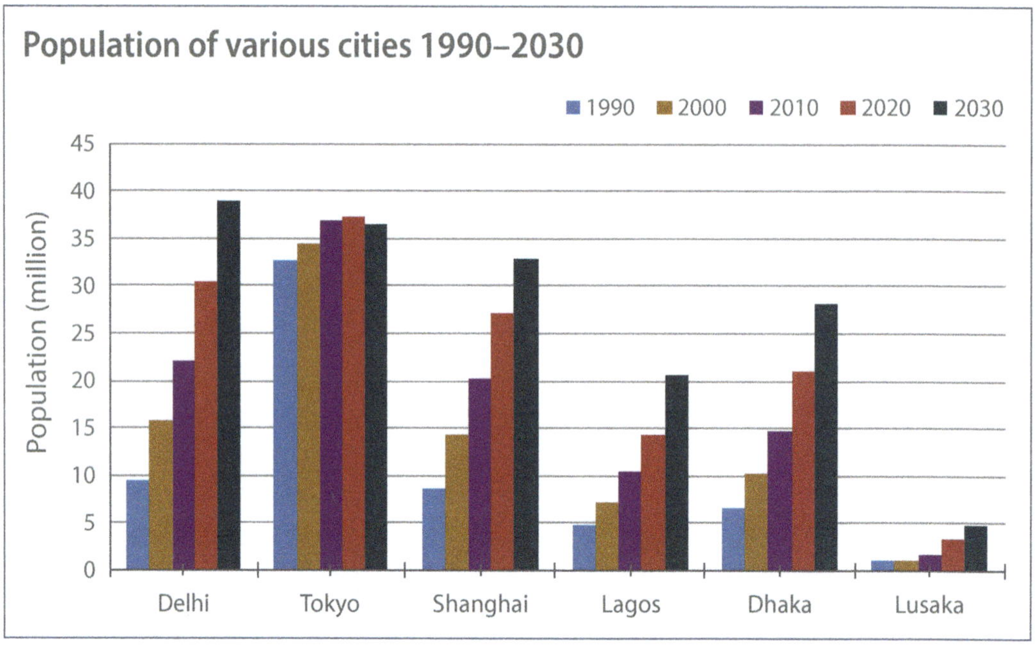

Figure 2.1: **Population trends in selected world cities**

Source: Data from http://worldpopulationreview.com/world-cities/ https://www.cia.gov/library/publications/the-world-factbook/fields/2219.html; https://en.wikipedia.org/wiki/Timeline_of_Lusaka#cite_note-UN1995-22; https://www.zamstats.gov.zm/phocadownload/Zambia%20Census%20Projection%202011%20-%20 2035.pdf (all accessed 04/03/2019)

Rural-urban migration: The process of moving from a rural area to an urban area.

Push factors: Factors making people want to leave an area.

Pull factors: Factors making people want to go to an area.

SECTION 2.1 PRACTICE

1. Study Figure 2.1.

   State the city that will have seen the greatest growth in population between 1990 and 2030.

   ....

2. Suggest why Tokyo's population is predicted to fall.

   ....

   ....

   ....

   ....

3. The population figures for 1990 and 2030 are shown below. Calculate which city will see the highest percentage growth rate between these dates.

CITY	1990	2030	% growth rate
Delhi	9,384,209	38,938,697	
Tokyo	32,530,003	36,573,799	
Shanghai	8,605,812	32,869,265	
Lagos	4,764,093	20,600,156	
Dhaka	6,620,697	28,075,660	
Lusaka	982,362	4,704,135	

Source: See Figure 2.1

4. Suggest two further **push** and **pull** factors leading to rural-urban migration.

   ....

   ....

   ....

   ....

5. Explain why LIC cities tend to have higher amounts of rural-urban migration than HIC cities.

   ....

   ....

   ....

   ....

2.2 Centripetal Population Movements in Post-industrial Cities

The industrial revolution began around 1750 in England and transformed society in Western Europe and North America. It is useful to have this in your mind, as it was not a global revolution.

Due to the large swathe of urbanisation occurring from the dawn of industrialisation in the mid-1700s, cities in high income countries tend to experience more gentrification than rural-urban migration. **Gentrification** refers mainly to the redevelopment of housing areas, particularly those in poor inner city areas. The subsequent change in demand for services by people of higher affluence can completely transform the district. To this end, gentrification is often linked into urban renewal. This can be witnessed in many cities around the world, from Glasgow to Rio de Janeiro (see section 2.5 for a case study on Glasgow and section 3.6 for one on Rio de Janeiro). Cape Town, Manilla, and Mexico City have also had elements of success in gentrifying neighbourhoods, indicating it is not confined to HIC cities.

Gentrification has arguably been most evident in North America, often resulting in the displacement of traditionally poorer ethnic groups. In Atlanta, Georgia, for example, rental prices have gone up by 28% since 2000 (Lartey, 2018).

There has been speculation that 'climate gentrification' could be an issue in the future as certain cities prepare for an influx of wealthier citizens from inhabitable cities. This could be considered a geographical 'possibility' in Arizona, read: https://www.theguardian.com/environment/2018/sep/25/climate-gentrification-phoenix-flagstaff-miami-rich-poor

CASE STUDY

Parkdale, Toronto

Parkdale is a region to the west of downtown Toronto and although still home to a large percentage of low-income housing in the south of the region, the northern part has been transformed and there has been an influx of younger, wealthier residents who can afford the increase in rent.

> *Gentrification is a beast with countless faces in Toronto. Many of them have shown themselves in Parkdale. The social, cultural and economic fabric of one of the city's last affordable neighbourhoods is being reshaped. And it's happening without meaningful conversation with those who currently call it home and will be forced out.*
>
> Barrington-Bush, 2018

Although the longer-term impacts of the Covid-19 pandemic on urban environments are yet to be determined, somewhat surprisingly gentrification actually increased in some cities, such as New York City: https://zora.medium.com/the-cities-where-gentrification-and-covid-19-collide-f672f44f0a9b

Also noted in the gentrification process:

- A development plan in 2018 offered zero units of affordable housing and encouraged more than 700 higher-income property owners.
- Original locally run Tibetan, Indian and Caribbean restaurants are being gradually replaced by chains such as 'Vegandale' to cater for the new clientele. The opening of the fifth chain in the area in 2018 led to protests (Krishnan, 2018).

A photo series dedicated by residents has been termed 'Vanishing Parkdale'. Information about this is at this link: https://www.toronto.com/community-story/7181882-photo-series-captures-impact-of-gentrification-in-parkdale/

Gentrification: The redevelopment of existing inner city areas, usually focusing on residential and characterised by an influx of more affluent residents.

SECTION 2.2 PRACTICE

1. Describe and explain the process of gentrification.

....

....

....

2. Referring to specific example(s), outline some pros and cons of the gentrification process.

....

....

....

3. Suggest the impact natural disasters and climate change could have for future gentrification proposals (the information link in the previous section is one source for further reading).

....

....

....

2.3 Centrifugal Population Movements

Centrifugal movements consist of **suburbanisation** and **counter-urbanisation**. Although these are similar, suburbanisation also refers to the physical growth of the city, perhaps into **greenfield sites** beyond the city boundaries. These processes tend to be more prevalent in high income cities, which have already past their peak of urbanisation and people are looking for improved environmental quality and cheaper housing. The processes are closely linked into urban sprawl. Los Angeles, USA has experienced suburbanisation for a number of years while Lisbon, Portugal is a good example of a city facing counter urbanisation. Ironically, a response to these processes can be urban renewal or gentrification to entice people back into inner city areas.

This free-to-view revision mind map looks at the causes and effects of suburbanisation in Los Angeles: https://getrevising.co.uk/diagrams/los-angeles-suburbanisation

This is a very interesting report on the likely increase in suburbanisation in European cities as the strive for urban environmental sustainability becomes even more apparent as a result of the Covid-19 pandemic: https://www.eea.europa.eu/publications/urban-sustainability-in-europe

Suburbanisation: The outward growth of urban development which may engulf surrounding villages and towns into a larger urban area.

Counter-urbanisation: The movement of people away from urban areas outside of the city.

Urban renewal/regeneration: Improving and redeveloping an urban area, often brownfield inner city sites or part of the CBD.

Greenfield site: An area of land that has not previously been built on.

SECTION 2.3 PRACTICE

1. Distinguish between counter-urbanisation and suburbanisation.

2. Based on the definitions of these terms, describe the main causes of centrifugal movements.

3. Suggest possible environmental and social consequences of centrifugal processes.

2.4 Urban System Growth and Infrastructure Improvements over Time

For urban areas to continue to develop and cater for increases in population, it is imperative that continual maintenance and upgrading of infrastructure takes place. Infrastructure can refer to transport, telecommunications, water and power supply, sanitation, and public institutions (like schools, hospitals, and prisons).

CASE STUDY

Rio de Janeiro, Brazil

Rio's site was attractive for Portuguese settlers in 1502, as it had the fertile soils in Guanabara Bay and was at the source of the River of January (Rio de Janeiro). Sugarcane, gold, and diamonds and the proximity to the Atlantic Ocean for trading allowed the area to develop economically. Following independence from Portugal in 1889, Rio became the capital city of Brazil and had a population of around 3 million.

You can read more about this here: https://www.riodejaneiro.com/v/history/

Although no longer the capital, today it is home to over 12 million people and is a dynamic, cosmopolitan megacity with a wide range of functions and extreme inequality. The World

Cup of 2014 and Olympic Games in 2016 (see section 3.6) have led to an abundance of spending in recent years, albeit with much controversy.

Transport

Rio has well-developed transport infrastructure, although the quality varies:

- The Metro opened on 5 March 1979, today covering 58 kilometres, serving 41 stations.
- The Galeão International Airport (open to passengers since 1952) links Rio to the rest of Brazil and has connections to 19 countries.
- The Santos Dumont National Airport (opened in the 1930s) is used now mainly for services to São Paulo and domestic flights.
- The Port of Rio is Brazil's third busiest port and the centre for cruise vessels and has been in use since the Portuguese first arrived.
- Buses are the main form of public transport, with 440 bus lines serving four million passengers a day. Although regarded as cheap and frequent, there have been countless protests over the quality of the services and the congestion on the roads.
- Rio has the oldest electric tramway in South America, the Santa Teresa Tram. Opened in 1877, it is mainly used by tourists today.
- A new $450m six-lane tramway was built for the 2016 Summer Olympic Games with a capacity of 285,000 passengers a day.
- It has an advanced number of expressways, reflecting the increased use of the car since 1990.

Telecommunications

The reach of media in Rio in 2016 was as follows:

- Free TV 86% (launched in 1950 for the wealthy elite);
- Internet 66% (up from 20% in 2000 and 47% in 2012);
- Radio 62% (launched in 1936, this figure has fluctuated since);
- Pay TV 48% (launched in 1995, this figure was only 20% in 2010).

Water and power

Brazil in general has benefited from rapid economic development and the impact of the **Millennium Development Goals**, as access to basic water supply rose from 94–97% between 2000–2015. The whole city is connected to a national electricity grid and power cuts are rare. Even in Rio's biggest favela, Rocinha (with unofficial estimates of up to 180,000 inhabitants), the majority have running water and 99% have electricity (Carneiro, 2014).

This website gives an overview of history and life in the *favelas* (Brazilian term for shanty housing) https://theculturetrip.com/south-america/brazil/articles/a-brief-history-of-rio-de-janeiros-favelas/

Millennium Development Goals: Superseded by the Sustainable Development Goals, these were 8 goals devised in the year 2000 by the UN to be achieved by 2015. They had various successes.

Sanitation

Although sanitation has improved in many parts of the city, it is suggested that 'one in three of the more than 10 million inhabitants of the greater metropolitan area live in places that have no connection to a sewerage system, and only about half of the city's waste is treated before entering waterways and eventually the ocean' (Vidal, 2016).

Summary and future infrastructure growth

Figure 2.2 shows investment in infrastructure as a percentage of GDP in Brazil since 1990. Trends show that levels fell below 2% in 2016, similar to that seen in the early 1990s and 2000s. This is highly frustrating for residents of Rio – other rapidly developing economies such as China and India invested between 5.5% and 7% of their GDP in the same year. The Olympic Games and World Cup bids promised to improve urban mobility, water, and sanitation. Rio's investments in mobility infrastructure have increased public-transport capacity, but the existing network remains inadequate for the metropolitan population. The water supply network is inefficient, with a leakage rate of around 50%. The biggest unfinished project is sanitation, a basic objective of the initial Olympic bid and the most ambitious of the United Nations Sustainable Development Goals. One of the targets was to clean up 80% of the effluents that reach Guanabara Bay. The actual proportion achieved was 55% (Osava, 2016).

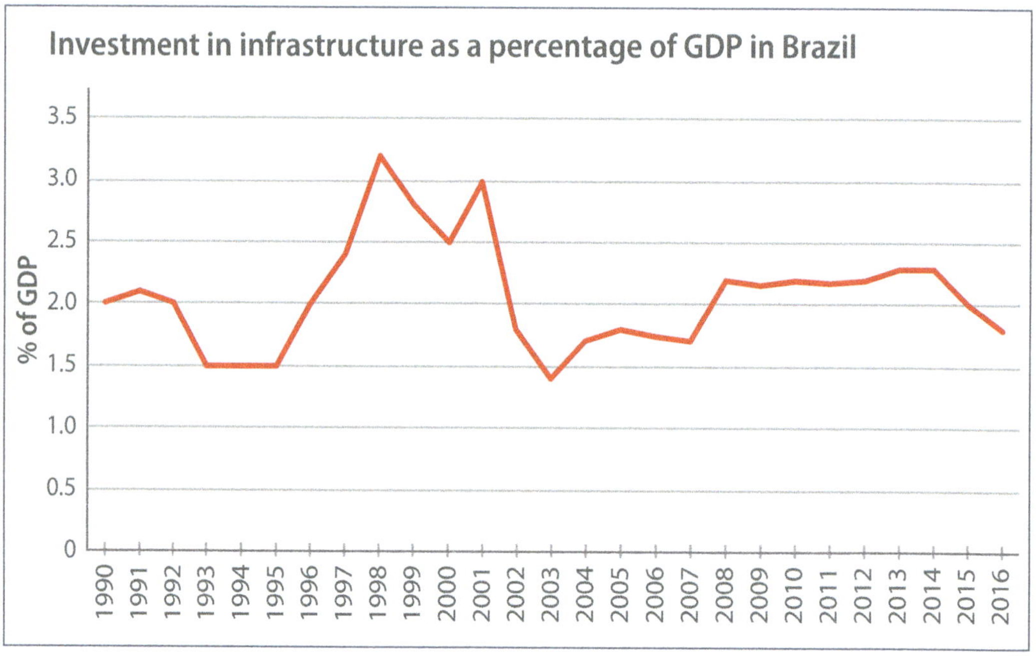

Figure 2.2: **Investment in infrastructure in Brazil**

Source: Data from Costa, 2018

SECTION 2.4 PRACTICE

1. Outline the reasons a well-developed infrastructure is vital for an urban environment to function.

....

....

....

2. Describe the prevalence of three types of infrastructure in a particular city.

....

....

....

3. Explain why disparities in infrastructure appear in cities like Rio de Janeiro.

....

....

....

....

2.5 The Causes and Consequences of Urban Deindustrialisation

Urban **deindustrialisation** was a characteristic of many European and North American cities during the twentieth century. As heavy secondary industries such as engineering became replaced by more tertiary and quaternary industries, many cities and their people began to lose their 'sense of identity' and suffered years of neglect and decay.

CASE STUDY 1

Glasgow

Glasgow's population peaked at 1.1 million between 1920 and 1950 (compared to around 620,000 in 2017) and was the 'second city of the British Empire', mainly due to shipbuilding on the River Clyde and its strong textile, iron, and steel industries. This did lead to overcrowding at this time: for example, the Gorbals district was said to house 85,000 inhabitants in only 2% of the city's territory in 1931. Stories of rats and street gangs were common in newspapers despite the overall success of the city.

The outbreak of WWII (in 1939) led to a temporary increase in employment in shipyard and heavy industry for those not enrolled in army service. However, bombings in the area and the subsequent decline of these industries meant Glasgow was facing a future of decline, increasing poverty, and crime. The city was over-reliant on shipbuilding and was unprepared for both the rapid development of other countries (such as China who could produce goods quicker and cheaper) and the unabated rise in technology. The car, the diesel locomotive, and then long-distance air liners meant many of Glasgow's traditional products were made redundant. A combination of poor planning in the city and lack of education in other spheres meant unemployment rose and productivity dropped, damaging economic output. By the late 1960s, there was little demand for passenger

Deindustrialisation: The process of areas ceasing or dramatically lowering production of heavy manufacturing industry, associated with many European and North American cities from the 1960s.

You can read more about this at the following links:

https://www.theglasgowstory.com/story/?id=TGSFO

https://www.bbc.com/news/uk-scotland-28419273

liners or ferries or steam locomotives. Skilled workers left the area from the 1970s, often for the fast-expanding North Sea Oil industry. Other sectors were also exposed: carpet makers depended on the shipyards for the bulk of their orders and textile spinners, weavers, and finishers succumbed to cheaper imports from abroad. By the 1990s only a handful of large manufacturing plants survived.

Unable to revive its industrial base, Glasgow had to look to the service sector to provide employment and has gone through large urban renewal and gentrification processes.

CASE STUDY 2

Chicago

Chicago emerged as a leading industrial centre in the USA in the early 1900s and was the second largest industrial city in the United States. This was due to the mass production of steel, clothing, meatpacking and agricultural equipment, as well as its strategic placement along canals and railroad lines.

Extra reading on this case study can be found at https://chryslervillageproject.wordpress.com/2015/10/29/deindustrialization-in-chicago-and-its-effects-1970-1990/

The Chrysler automotive plant was designed to mass produce airplane engines for B-29 bombers, and at its peak in early 1945 it employed over 30,000 workers. The plant was heavily impacted by the process of deindustrialisation due to the end of the war, the rise of the service economy, and latterly **globalisation**. The number of employees fell to just over 2,500 by December 1945. There was a continued decline in employment between 1970 and 1990 (the wider Great Lake Region lost over a million manufacturing jobs) as large businesses began **outsourcing** production to low-wage countries. Furthermore, technological advancements allowed factories to produce the same amount of goods or more with a smaller workforce and more spending was continuing to be directed to the service sector.

Chicago, prior to deindustrialisation, was a model of urban growth that focused its energies toward mass production of industrial goods. However, of the nearly one million **blue-collar workers** that were employed in 1979, over 30% had lost their jobs by 1990.

In the 'Clearing' neighbourhood alone, the number of workers employed in manufacturing fell from close to 50,000 in the 1950s to around 19,000 by 1984, and nearly half of its manufacturing firms closed or moved due to deindustrialisation. Overall, the population of Chicago has declined from its peak of over 3.2 million in 1950 to just over 2.7 million in 2017.

Globalisation: The increasing interconnectivity and interdependence of nations worldwide.

Outsourcing: Where a company hires another company to be responsible for existing activity, often in a different country.

Blue-collar worker: Workers who are usually involved in manual labour (as opposed to **white-collar workers** who are involved in administration of office work).

SECTION 2.5 PRACTICE

1. Briefly outline the meaning of the term 'deindustrialisation'.

....

....

2. Discuss the main causes of deindustrialisation in:

   (a) Glasgow

....

....

....

   (b) Chicago

....

....

....

3. Complete the table below for consequences of deindustrialisation.

	SOCIAL	ECONOMIC	DEMOGRAPHIC
Glasgow			
Chicago			

4. Examine the extent to which the social consequences of deindustrialisation are worse than the economic.

....

....

....

Chapter 3: Urban Environmental and Social Stress

Overview

As cities and their infrastructure continue to develop it is inevitable that certain environmental and social **stresses** emerge. These can range from general pollution problems to crime to various land use conflicts. Some of the causes, impacts and mitigation attempts are covered in this chapter.

KEY QUESTION

What are the varying powers of different **stakeholders** in relation to the incidence and management of urban stresses?

Geographic knowledge and understanding points

- The causes and management of microclimates in urban areas
- Urban stresses of air pollution and traffic congestion
- Land use conflicts
- Patterns of crime and deprivation

3.1 Urban Microclimates

An **urban microclimate** refers to the different climate an urban area, or a part of an urban area, can have in comparison to close-by, more rural areas. The temperature profile of the CBD is often referred to as an '**urban heat island**' because it can have temperatures of up to 2–4°C higher than surrounding land use zones.

Stakeholder: A person or group with an interest or concern in something.

Urban heat island: A metropolitan area that's a lot warmer than the rural areas surrounding it.

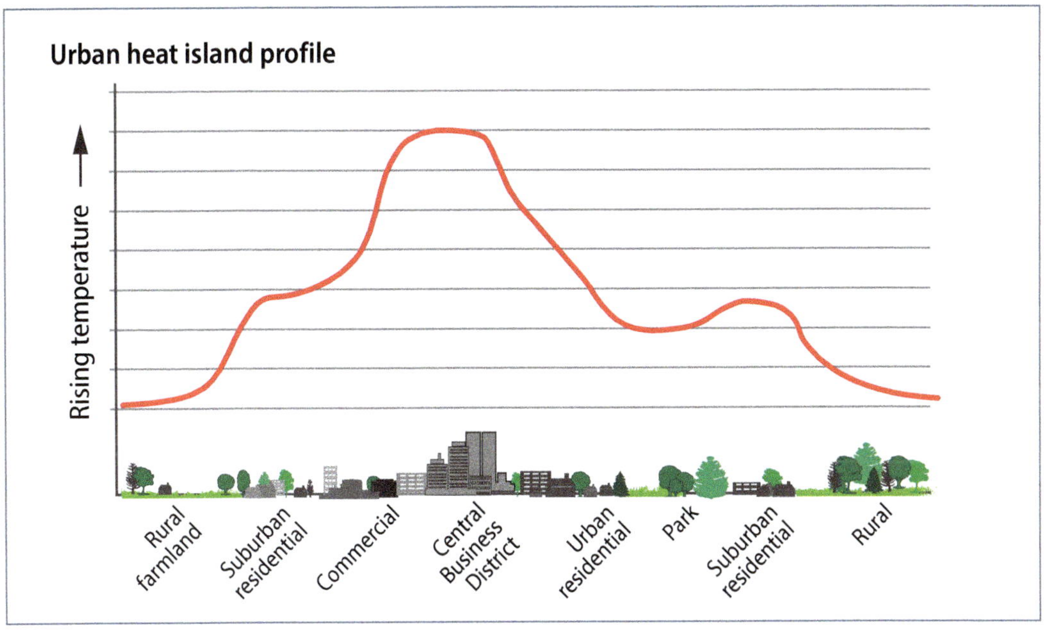

Figure 3.1: **Urban heat island**

Figure 3.1 shows a general urban heat island profile with the CBD notably warmer than surrounding rural areas. The reasons for this are:

- Higher absorption of **long-wave radiation** and energy from the sun due to darker, asphalt surfaces. This also lowers **albedo** levels.
- Higher air pollution results in more clouds trapping heat. (see section 3.2 on air pollution)
- Higher population density and the use of energy in buildings during the day traps in the heat at night.
- Tall buildings funnel wind causing a **canyon effect**, obstructing rural air flows.

Urban areas also tend to be drier than rural areas because of the lack of green space and a dominance of **impervious** surfaces. Urban drainage systems also quickly remove water from the urban surface and thus the atmosphere around the city warms more quickly/ effectively.

As the unabated rise of global temperatures continues, urban planners have significant challenges in managing and mitigating the continuing threats of further increases in the microclimate. These include, but are not limited to, rising sea levels, more extreme weather events, and increased air pollution.

Long-wave radiation: The energy radiating from the Earth as infrared radiation.

Albedo: The total amount of reflection of solar radiation.

Canyon effect: When tall buildings flank both sides of city streets it can artificially create a canyon-like effect. It can affect temperature, light, wind, and air quality.

Impervious: Not allowing water to pass through.

SECTION 3.1 PRACTICE

1. Briefly explain how an 'urban heat island' may be created.

....

....

....

2. Draw an annotated diagram to show how an urban microclimate can form, using key terms from the previous page.

3.2 Air Pollution Patterns

Air pollution has been a growing concern in the twenty-first century and is exacerbated in urban places where high levels of population, consumption, traffic, and industry can pose great health risks. While many cities have made progress in attempting to lower their air pollution levels an EU report from September 2018 stated that 'Air pollution is the "biggest environmental health risk" in Europe and governments are failing to tackle the crisis that causes 1,000 early deaths a day' (Nelson, 2018).

However, it is in India where levels are of the greatest concern. A study released by the WHO in 2018 concluded that India contained 9 of the 10 most polluted cities on the planet, with Delhi at the top.

CASE STUDY

Delhi

With an estimated population of over 28 million in 2018 (World Population Review) and a growth rate of over 3% a year, Delhi has a heavy presence of PM_{10} (particular matter): 292 micrograms per cubic metre. The annual mean safe limit set by the WHO is 60. On occasion it has gone dramatically higher: in November 2017, in a period known as the 'Great Smog of Delhi,' it hit 999 micrograms per cubic metre. The following month in December during a test cricket match in the city against Sri Lanka, several players felt breathing problems and vomited both in the rest rooms and in the field and had to use face masks until the match was stopped (Ananthanarayanan N. et al., 2017).

Figure 3.2: **An overview of how Delhi's air quality compares to other major global cities**

Source: McCarthy. 2018. Delhi Has The Worst Air Quality In The World. https://www.statista.com/chart/13814/delhi-has-the-worst-air-quality-in-the-world/ (accessed 04/03/2019) CC BY-ND

Impact on people

From reports of toxic foam on the holy River Yamuna to 'pollution migrants' the problem is widespread across the city. 'Many of my elderly patients leave Delhi in the colder months and live with their children or family members in other cities,' says Dr Vivek Nangia, head of Fortis Lung Centre in the city (Duttagupta, 2018). Overall poor air quality is a cause of reduced lung capacity, headaches, sore throats, coughs, fatigue, lung cancer, and early death. Older people, children, pregnant women, and cancer and diabetes patients are

Terms like 'pollution migrant' are not in the wide domain and can be used to emphasise geographical possibilities. As pollution worsens, people feel the need to move from their homes.

more vulnerable to complications and it is estimated that 2.2 million children in Delhi (around 50%) have irreversible lung damage due to the poor quality of the air. In addition, research shows that pollution can lower children's immune system and increase the risks of cancer, epilepsy, diabetes, and even adult-onset diseases like multiple sclerosis (Harris 2015).

Another prominent doctor in the city, Arvind Kumar stated in a 2018 report on NDTV in Delhi that 'I have seen a change in the colour of lungs over the past 30 years that I have been operating. Earlier, I used to see black deposits only in smokers and others would have pink lungs. But, nowadays, I only see black lungs. Even teenagers have black spots on their lungs. This is frightening.'

SECTION 3.2 PRACTICE

1. Referring to a specific city, explain the impact of air pollution on the people.

....

....

....

2. Study Figure 3.2. Describe the distribution of the selected cities.

....

....

....

3. Suggest why Delhi and New York have such drastic differences in their concentration of PM_{10} particles.

....

....

....

....

3.3 Traffic Congestion and its Management

CASE STUDY

Brussels

Although not witnessing the levels of air pollution as residents of Delhi, Brussels frequently suffers from levels of nitrogen oxide (NO_2) well above the EU limit of 40 micrograms per cubic metre (Hope, 2018). Furthermore, a Greenpeace study found Belgium had one of the highest concentrations of NO_2 in the world, as seen from space. This is particularly centred on the northern city of Antwerp and its port industry, and the high levels of diesel cars and congestion in Brussels.

Traffic congestion has been an issue in Brussels for a number of years, with reports of a 40% increase in the number of cars driving in to the city during peak hours between 1986 and 2002 (Expatica, 2004). In 2012 and 2013 Brussels was Europe's most congested city according to the Inrix scorecard and this led to the unwanted title of 'traffic-jam capital of the world' in September 2013.

You can read this news report at http://www.flanderstoday.eu/current-affairs/brussels-traffic-jam-capital-world

Some of the reasons suggested for the high rates of traffic congestion are the large number of company cars (partly due to the high tax rates in the country), high rates of suburbanisation, a lack of park-and-ride facilities and relatively cheap and abundant inner city parking. Cycle infrastructure is still very poor in many suburban areas and in the hilly inner city areas.

However, there is evidence that some successful management strategies have been put in place, as shown in Figure 3.3 and Table 3.1.

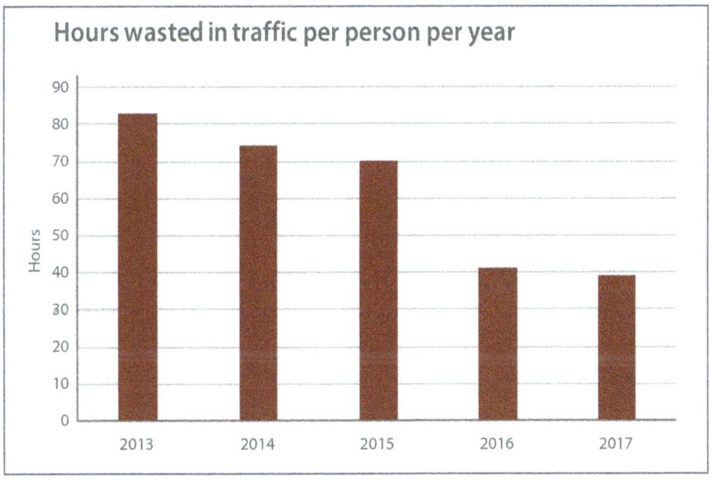

Figure 3.3: **Average hours wasted in traffic per person per year in Brussels**
Source: Data from INRIX. http://inrix.com/scorecard/ (accessed 04/03/2019)

City	Hours in congestion (2015)	Hours in congestion (2017)	Difference
London	101	74	–27
Brussels	70	39	–31
Moscow	57	91	+34
Munich	53	51	–2
Manchester	51	39	–12
Paris	45	69	+24

Table 3.1: **2015 and 2017 selected European cities – hours spent in congestion**
Source: Data from INRIX. http://inrix.com/scorecard/ (accessed 04/03/2019)

By 2017 the city had dropped to 20th position in European cities monitored for hours spent in congestion. Although these figures are slightly skewed by the addition of more cities (1,360 cities in 38 countries worldwide), it suggests that the number of hours wasted in traffic congestion in Brussels dropped by more than a half in five years. Meanwhile Paris, for example, has reportedly risen from 45 hours in 2015 to 69 in 2017.

A key contributing factor is investments in suburban rail services in and around the city to help ease gridlock. The reduction could also partly be attributed to the pedestrianisation of two major squares by the Bourse Stock Exchange in July 2015. Although progress has been slow on this, the scheme was dubbed the 'largest car-free zone in Europe after Venice'. There have also been an extra 3,000 metres of cycle lanes developed around the area (Banks, 2017).

Other methods to reduce congestion and encourage alternatives to driving in the city include:

- the 'Villo!' rental bike scheme (see Figure 3.4) and 'Uber E-Bikes';
- 'Mobility week': the idea here is to reduce the density of traffic and give back public spaces to Brussels residents by encouraging car sharing and public transport;
- car-free days;
- various electric scooters (such as Lime and Dott) have been placed strategically around the city, similar to the Villo! and Uber E-Bike schemes;
- extension of tram lines;
- development of electric buses;
- potential surcharge for diesel cars.

Despite these successes, a poll from the Brussels Bulletin in December 2018 found that 91% of respondents still consider traffic a 'major problem' and demonstrations for pedestrian safety from heavy traffic were still taking place in September 2021. Perhaps the ambition shown by the city of Madrid, Spain in banning all polluting vehicles from their CBD in the same month may be the only way forward in eliminating congestion and associated pollution.

Figure 3.4: **'Villo!' scheme in Brussels to discourage car use**

SECTION 3.3 PRACTICE

1. Describe the main causes of traffic congestion.

2. Referring to a specific city, evaluate possible management responses to traffic congestion.

3.4 Contested Land Use Changes

Section 1.2 outlined urban land uses and potential models to demonstrate these. Land use changes can cause conflict between different groups of people, such as government officials, local residents, employers, or other stakeholders. The result of these can be slum clearances such as in the Barra de Tijuca *favela* in Rio de Janeiro (the second case study in this section) and the increasing urbanisation and the depletion of green land such as in Kumasi city in Ghana (the first case study below). Another example of land use conflict is in the eastern neighbourhood of Stratford, London, which underwent large-scale urban renewal and gentrification in preparation for the Olympic Games in 2012. Furthermore, official statistics show that more than 50,000 cases of land use dispute took place in the first decade of this century in China. This was during their rapid urbanisation process with reports of people being 'prisoned and tortured trying to protect their house and lands' (Magsi, 2017).

AFFECTED NEIGHBOURHOOD: CASE STUDY 1

Danyame neighbourhood, Kumasi city, Ghana

Census data show that Kumasi's population (just over 2 million in 2010) maintained a growth rate of over 5% per annum between 1984 and 2010, making it the fastest growing major city in Ghana. This resulted in the expansion of the built-up area and the conversion of farmlands to urban uses. The total built-up area of the city almost quintupled from approximately 98 square kilometres in 1972 to 478 square kilometres in 2012 – a growth rate of 9.5 square kilometres per annum. This was due to both rural–urban migration and mergers with nearby semi-urban communities such as Danyame. The combined effect of these two processes (namely suburbanisation and urban sprawl) is that the built-up area has now expanded well beyond the administrative boundaries of the Kumasi Metropolis (Yaw Oduro *et al*, 2014).

Due to this rapid urbanisation the inevitable slums have appeared around the city. According to the 'Modern Ghana' website, one known as 'Abinkyi slum' was bulldozed in 2010, evicting approximately 2,000 people.

A city plan of 1945 made conditions to protect many of the city's green spaces in and around Danyame, but these have since been encroached upon. For example, a nature reserve and public park in the initial planning map do not exist today. Despite the 'Keep Kumasi Clean and Green' project launched in 2018, a vast area of green vegetation was destroyed by an unknown developer in March 2020. The depletion of these green spaces appears to have multifaceted causes and has created many land use conflicts. Some of the reasons for this are:

More information on this can be found here https://www.ghanaweb.com/GhanaHomePage/NewsArchive/We-don-t-know-who-destroyed-Danyame-vegetation-Kumasi-Mayor-884371

- poor enforcement of maintaining green space by city authorities and too many different bodies constituting the planning (e.g. Town and Country Planning Department, Development Control Unit, and Development Planning Unit);
- political interference and lack of finance;
- ownership of green land is confusing – there are 'vested' lands owned by the government, 'stool lands' owned by chiefs (around 60%), and 'public lands' which are acquired by government for public interests (e.g. railways and sanitary sites);
- lack of enthusiasm by locals towards the preservation of green space as they use spaces to sell and perform general commercial activities;
- top priorities for Kumasi city authorities are education, health, and commercial activities;
- no specific 'green plan' or strategy today;
- the poor maintenance of green spaces encourages developers and individuals (who see them lying idle) to encroach them for their personal interests (Collins, 2014).

AFFECTED NEIGHBOURHOOD: CASE STUDY 2

Barra de Tijuca, Rio de Janeiro, Brazil

The development of infrastructure in Rio was outlined in section 2.4, but one of the most controversial aspects of the many development projects for the 2016 Olympic Games was the conflict that occurred in the western suburb of Barra de Tijuca. Overall, in the 2015 book, *SMH 2016: Evictions in Olympic Rio de Janeiro*, it was calculated that more than 60,000 people in the city lost their homes between 2009 and 2013, and it argued that there was no more room for the poor. Critics say that infrastructure built for the Olympics, such as three new bus rapid transit (BRT) lines, has come at a high price for some poorer residents and only built for short-term benefit.

These images clearly show the transformation of Rio following the Olympic games: http://graphics.wsj.com/rio-city/

Figure 3.5: **Copacabana beach in Rio de Janeiro being prepared to welcome tourists for the World Cup and Olympic Games – both led to land use conflicts**

Source: Kenny Chung. http://kennychung.net/photo/rio-de-janeiro/ipanema-beach/ (accessed 10/4/2019). CC BY-ND 3.0

In Barra, a reported 275 of 800 families were ordered to leave the 'Vila Autódromo' *favela* in the area. While some evictions cleared space to build or expand infrastructure, activists and residents say the city's only purpose was to boost land values in the Olympic Park, used for private real-estate development after the Games (Phillips, 2016).

The new Olympic Golf course was also developed here. Although many residents refused to leave, generally they were relocated to other low-income housing projects, although some were transferred as far as 40 kilometres away (Cuadros, 2016).

SECTION 3.4 PRACTICE

1. Analyse the factors that can lead to 'contested land use.'

....

....

....

2. Referring to two contrasting examples, explain how conflicts in land use can develop.

....

....

....

3. Discuss the impacts of:

   (a) The depletion of green space

....

....

....

   (b) The demolition of slum housing

....

....

....

3.5 Managing Social Deprivation and Geographic Patterns of Crime

The incidence of deprivation and poverty were covered in sections 1.6 and 1.7 and these are often associated with higher levels of crime.

CASE STUDY

London

Managing deprivation is a major challenge for the city of London and is directly related to the high levels of violent crime, identified by the city's mayor (Sadiq Khan) as a problem that 'could take a generation to solve' following three knife murders in five days in October 2018. That year (up until November) there were 118 homicides, including 73 stabbings and 12 shootings, compared to 116 for the whole of 2017. Furthermore, the terror incident on London Bridge in November 2019, in which five people were stabbed, brought these issues further into the media spotlight. According to City Hall figures police coverage in London was at its lowest rate since 1998 with 3.3 police officers for every thousand Londoners – 19% lower than in 2010.

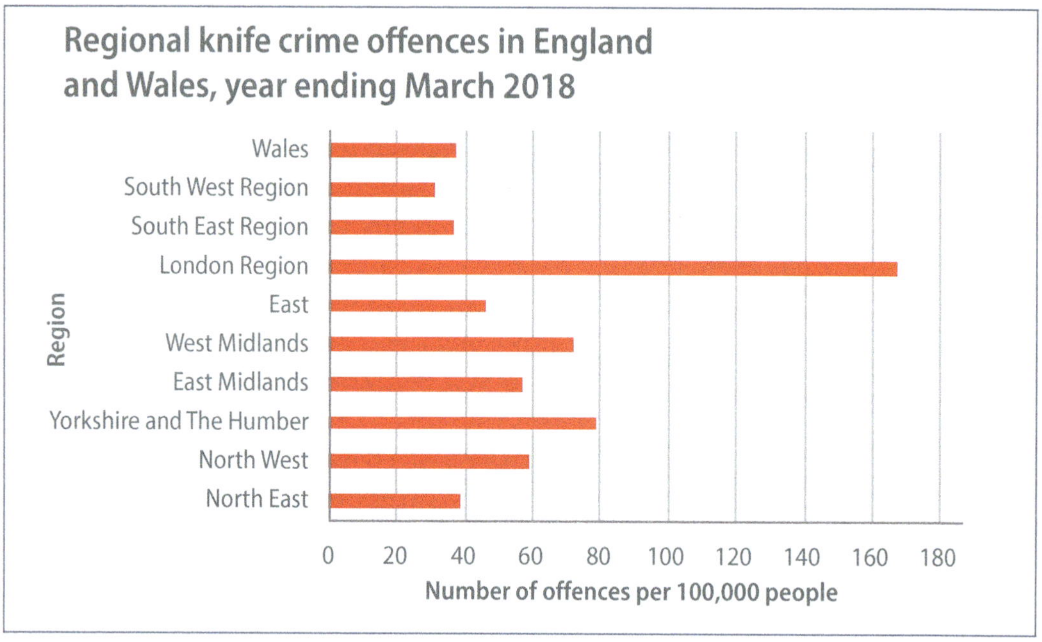

Figure 3.6: **Knife crime in England and Wales, 2018**

Source: Data from Office for National Statistics https://www.ons.gov.uk/peoplepopulationandcommunity/crimeandjustice/datasets/offencesinvolvingtheuseofweaponsdatatables (accessed 04/03/2019)

Figure 3.6 illustrates the extent of knife crime in London compared to other regions in England and Wales, and Figure 3.7 identifies contrasting crime rates in different London boroughs.

London has seen some success in lowering social deprivation, such as in the borough of Newham in the East. In 2015 the borough went from the 2nd most deprived local authority in England to the 25th. Newham Council fought hard for strong local employment quotas in the 2012 London Olympics regeneration projects and revamped their job scheme (called 'Workplace'). This involved hiring managers to visit the offices of major local employers and identify and hire Newham residents. The result was 80% of clients that were registered finding employment, compared to 2% with the old method, where clients were just told about job openings and then had to apply themselves (Presser, 2016). In terms of crime, Newham was not the worst borough at around 90 per 100,000 people in 2015, but it was still in the top 10 and has reportedly seen increased levels since.

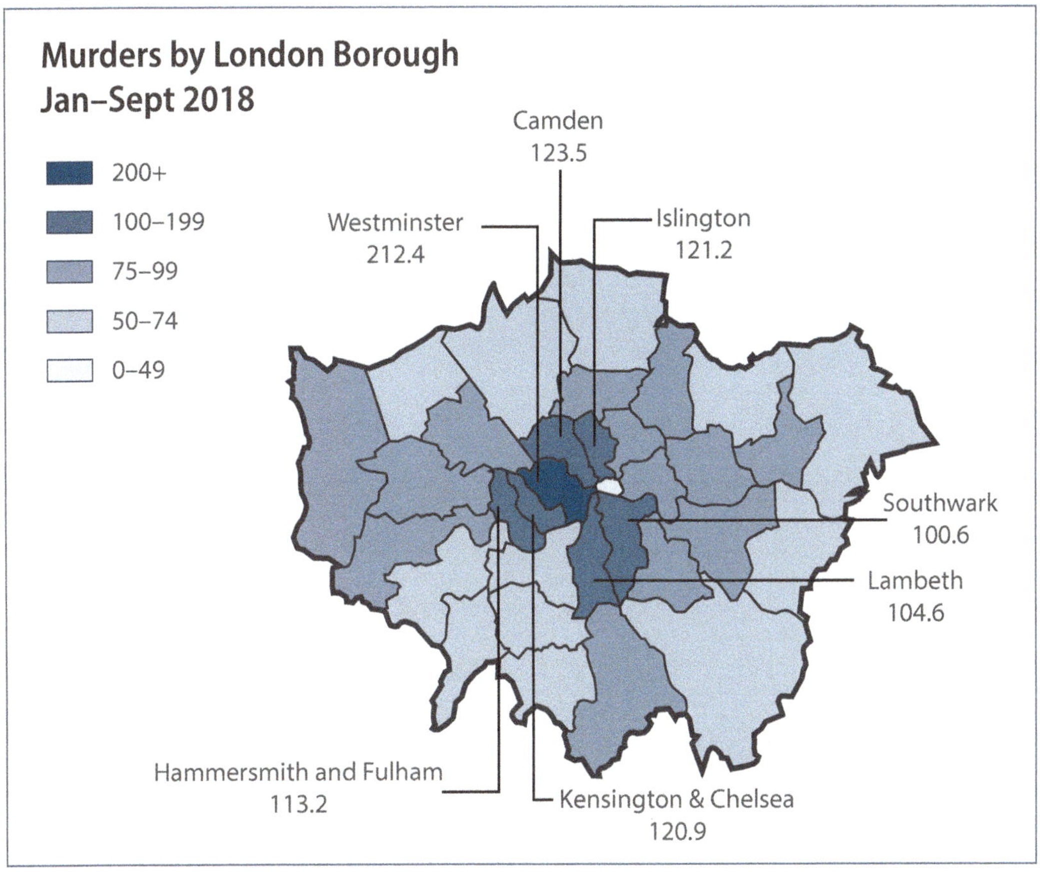

Figure 3.7: **Crime rates by London Borough (per 1000 of the population), 2015**

Source: Data from https://data.london.gov.uk/dataset/london-borough-profiles (accessed 04/03/2019)

One method of managing urban crime has been developed by the Scottish Government who have managed to halve the homicide rate between 2007 and 2017, a third of which was from the city of Glasgow. By viewing violence as a disease, it aims to diagnose the problem and treat its cause. This has seen police officers working with teachers and social and health workers to collate and share knowledge of people involved in gangs. Although Glasgow and London are difficult to compare in terms of size, population and the way crime statistics are reported, it has been suggested that London could adopt this method of managing crime (Mackintosh, 2018). Further suggestions for managing crime are included in section 4.5.

SECTION 3.5 PRACTICE

1. Analyse the incidence of crime in London.

2. Describe the geographic pattern and links between deprivation and crime rates in London (see Figure 1.15 on page 19 and Figure 3.7).

3. Suggest ways in which urban crime levels can be reduced.

Chapter 4: Building Sustainable Urban Systems for the Future

Overview

With the unabated global rise of urban populations projected to continue, the sustainability of urban settlements is crucial. This needs to be thought of in a multidimensional manner with social, economic, and environmental factors all of great importance. However, the power that comes from political governments is equally vital, and they have a huge part to play in ensuring that future generations can reside in safe, clean urban places. Recent developments have seen different titles given to city planning. Whether they be considered 'resilient', 'eco' or 'smart', they all have the ultimate aim of achieving sustainability in urban environments. The sudden, unexpected outbreak of Covid-19 and subsequent urban lockdowns in 2020 and beyond add an interesting dimension to this topic and it will be fascinating to see how cities bounce back. Will future plans for sustainability incorporate preparation for future pandemics? Have lockdowns changed the face of city life forever? These questions will be answered as we move through the 2020s.

Sustainable Development Goals 9 and 11 directly relate to the issues in this chapter.

KEY QUESTION

What are the future possibilities for the sustainable management of urban systems?

Geographic knowledge and understanding points

- Projections for future urban growth
- Strategies to manage climate change and geopolitical tensions
- Designing an 'eco-city' and managing the urban ecological footprint
- The impact of technology and the development of smart cities

4.1 Urban Growth Projections for 2050

This fantastic resource is updated annually and is available to download here: https://population.un.org/wup/

Projections for future population growth are made frequently and vary in terms of their validity. It is generally accepted by studying the UN's World Urbanisation Prospects that we will have approximately 9.8 billion people in 2050 and that 68% of these people (6.7b) will live in urban areas, compared to around 55% (4.2b) today. That equates to a further 2.5 billion **urbanites** to cater for, with close to 90% of this increase taking place in Asia and Africa. To put this in perspective, in 1950 there were 751 million people considered urbanites. Figures 4.1 and 4.2 show some population statistics and projections.

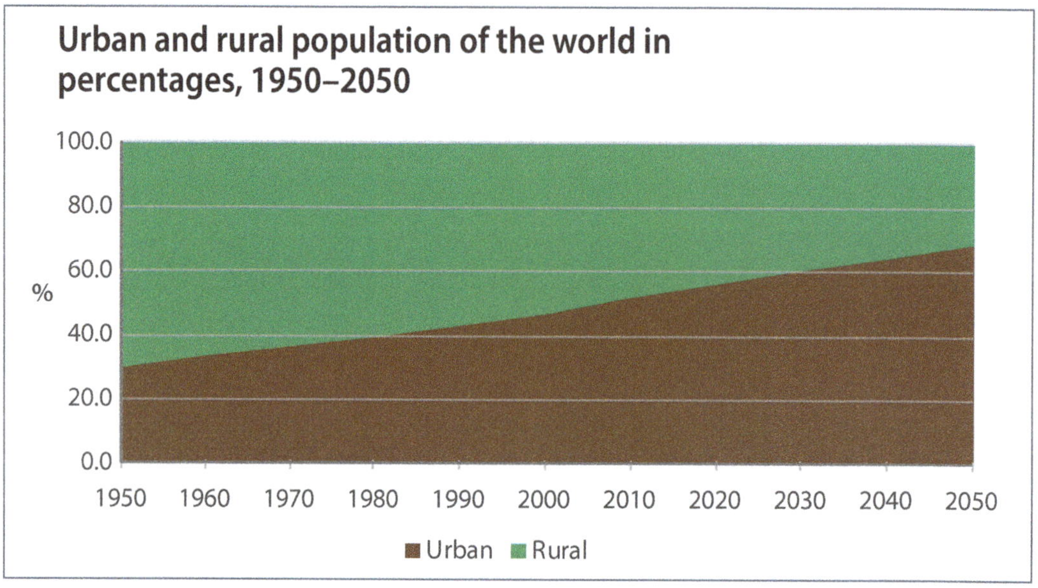

Figure 4.1: **Urban population of the world 1950-2050**

Source: Data from United Nations, Department of Economic and Social Affairs, Population Division. 2018. World Urbanization Prospects: The 2018 Revision. https://population.un.org/wup/ (accessed 04/03/2019)

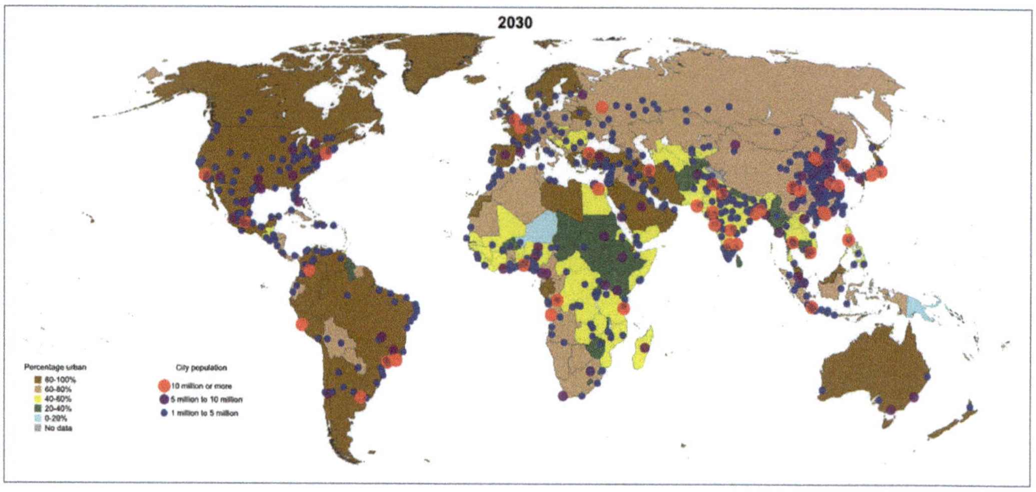

Figure 4.2: **Projected percentage urban and cities over 1 million, 2030**

Source: United Nations, DESA, Population Division. 2018. https://population.un.org/wup/Maps/ (accessed 04/03/2019). CC-BY-3.0 IGO

Urbanite: A person who resides in an urban area.

Together India, China, and Nigeria will account for 35% of the projected growth of the world's urban population between 2018 and 2050 with an extra 416 million urbanites in India, 255 million in China, and 189 million in Nigeria.

The most urbanised regions are Northern America (with 82% of its population living in urban areas in 2018), Latin America and the Caribbean (81%), Europe (74%) and Oceania (68%). The level of urbanization in Asia is now approaching 50%. In contrast, Africa remains mostly rural, with 43% of its population living in urban areas. See Figure 4.3 for a comparison of continental urban growth rates.

SECTION 4.1 PRACTICE

1. Using the text and figures 4.1 and 4.2, describe the patterns and trends of urban growth from 1950-2050.

....

....

....

....

2. Suggest why urban environments in Africa are projected to grow the most.

....

....

....

3. Outline reasons why urban places in HICs are predicted to grow at a far slower pace.

....

....

....

....

4.2 Trends of Rural-urban Migration

Earlier sections on megacities (1.3) and centripetal movements (2.1) outlined some of the reasons for, and the scale of, rural-urban migration. Figures 4.3, 4.4 and Table 4.1 show urban growth rates and projections for continents, countries, and cities respectively.

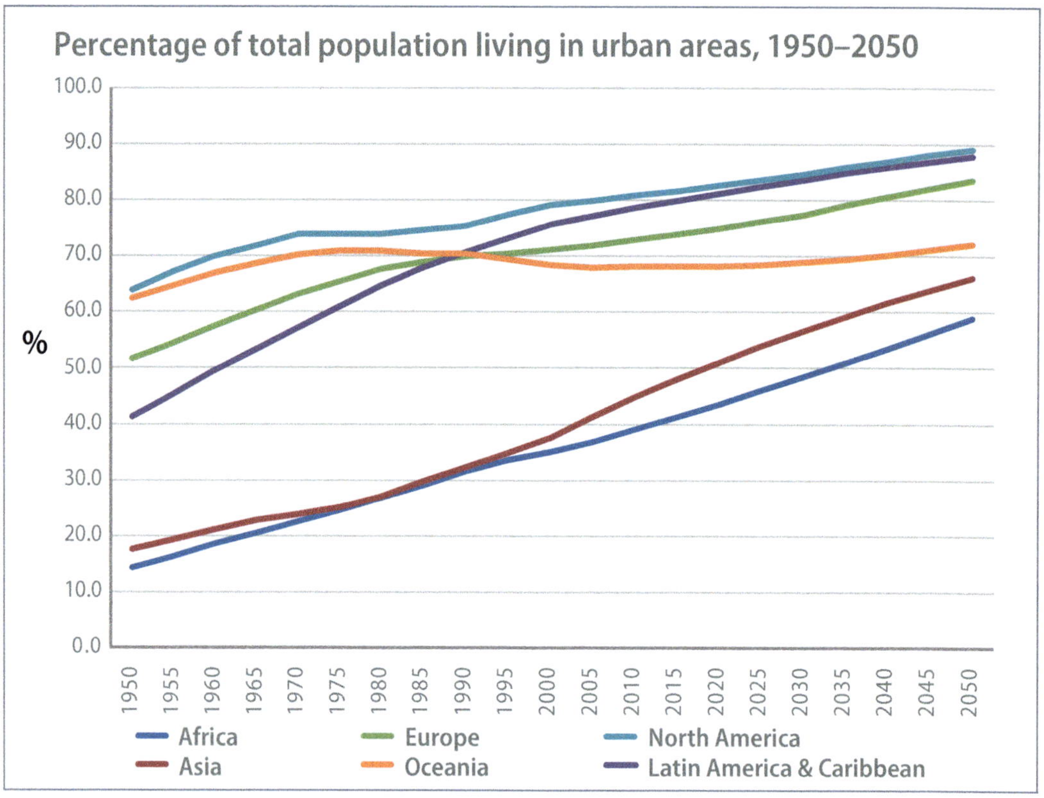

Figure 4.3: **Continental urban growth rates 1950–2050**

Source: Data from United Nations, Department of Economic and Social Affairs, Population Division. 2018. World Urbanization Prospects: The 2018 Revision. https://population.un.org/wup/ (accessed 04/03/2019)

Country	Urban population growth 1960 (annual) %	Urban population growth 1990 (annual) %	Urban population growth 2015 (annual) %	Economic class
Niger	4.54	3.6	5.44	LIC
Zambia	7.35	2.64	4.18	LMIC
Bangladesh	4.67	4.92	3.44	LMIC
India	2.34	3.03	2.38	LMIC
Brazil	5.22	2.8	1.15	HMIC
United Kingdom	0.28	0.23	1.09	HIC
Poland	5.33	0.47	-0.11	HIC

Table 4.1: **Selected countries and their urban population growth 1960–2015**

Source: Data from The World Bank, Health Nutrition and Population Statistics. https://datacatalog.worldbank.org/dataset/health-nutrition-and-population-statistics (accessed 04/03/2019). CC-BY 4.0

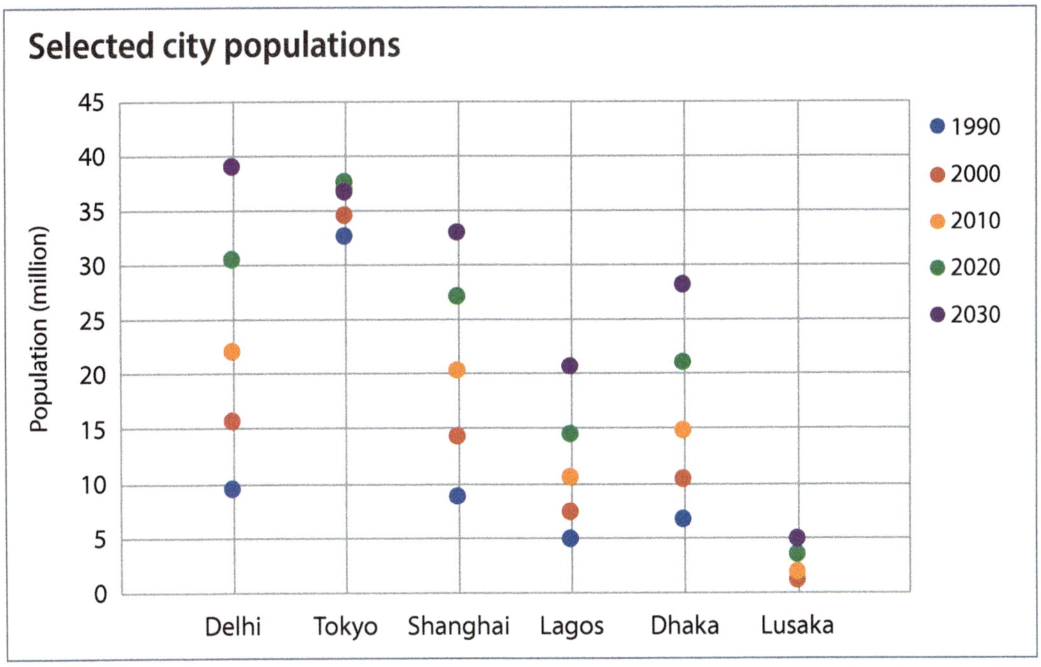

Figure 4.4: **Selected cities and their populations 1990–2030**

Source: Data from: http://worldpopulationreview.com/world-cities/ ; https://www.cia.gov/library/publications/the-world-factbook/fields/2219.html; https://en.wikipedia.org/wiki/Timeline_of_Lusaka#cite_note-UN1995-22; https://www.zamstats.gov.zm/phocadownload/Zambia%20Census%20Projection%202011%20-%202035.pdf (all accessed 04/03/2019)

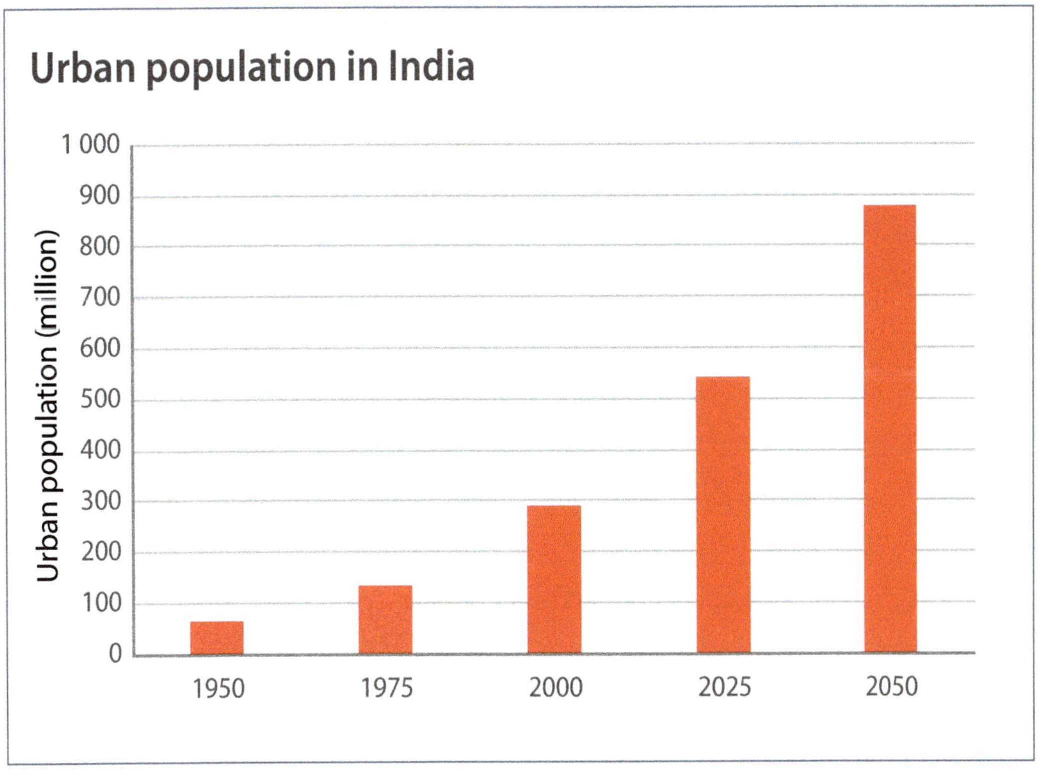

Figure 4.5: **Urban population in India**

Source: Data from United Nations, Department of Economic and Social Affairs, Population Division. 2018. World Urbanization Prospects: The 2018 Revision. https://population.un.org/wup/ (accessed 04/03/2019)

SECTION 4.2 PRACTICE

1. Study Figure 4.3 on page 48. Examine the consequences of urban growth for different continents from the year 2000.

2. Study Table 4.1 on page 48. Describe the trends in urban growth rates from 1960–2015.

3. Suggest three reasons for these trends.

4. Explain why it may be more accurate to look at urban growth rates rather than the population size of cities.

5. Study Figure 4.4 on page 49.

   (a) Identify the number of megacities in the year 2000.

   (b) Estimate the population of Shanghai in 2020.

   (c) Calculate the population growth of Tokyo between 1990 and 2020.

6. Study Figure 4.5 on page 49. Describe the growth in urban population in India between 1950 and 2050.

4.3 Changing Urban Population Sizes and Structures

Population sizes and structures are often visualised in population pyramids. Figure 4.6 shows two pyramids for a city, one in the year 2000 and one in the year 2050. This shows an overall increase in population of the city, but a healthier structure. The year 2000 has a very youthful population structure but due to economic development, lowering birth rates and improvements in healthcare the dynamic has completely changed by 2050. It is evident the city has experienced a **demographic dividend** during this period and will likely face issues of an ageing population further in the future.

SECTION 4.3 PRACTICE

1. Compare and contrast the population size and structure of the city in Figure 4.6 on page 52 in the years 2000 and 2050.

2. Suggest reasons for your answers to question 1.

Demographic dividend: This refers to a 'bulge' in the adult, economically active population and is generally considered a favourable economic position for a city.

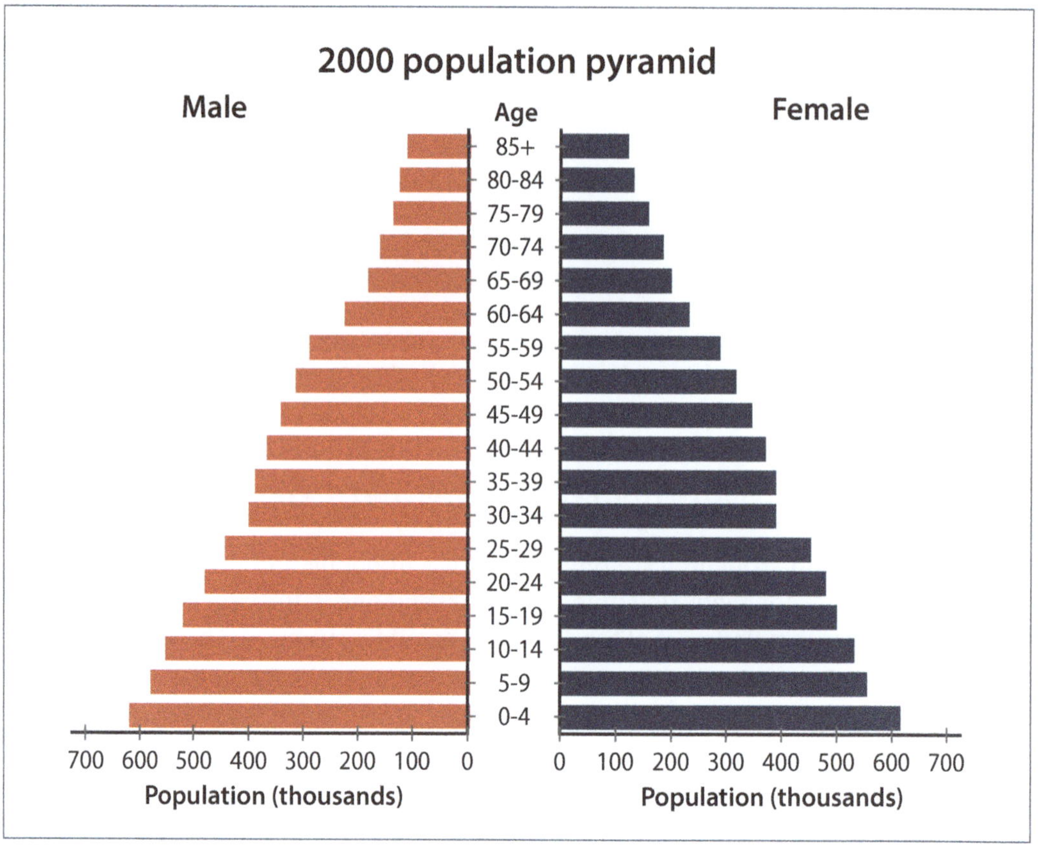

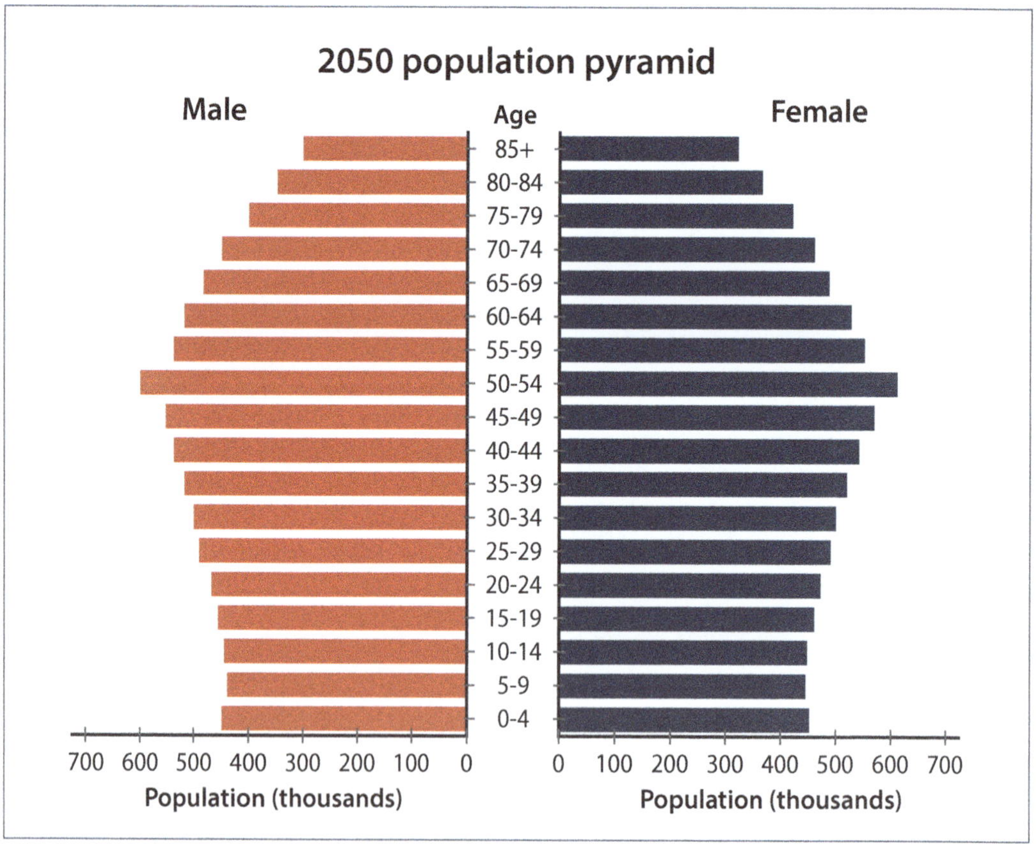

Figure 4.6: **Population pyramids of a city in 2000 and 2050**

4.4 Resilient City Design: Climatic Control

Climatic and geo-political urban stresses are only likely to intensify in the coming years; therefore, cities are attempting to become more **resilient**. For example, a resilient city may focus on adapting to future climate change by reducing traffic congestion and associated air pollution. The '100 resilient cities network' has been established as cities attempt to mitigate the risks.

CASE STUDY

Melbourne

Australia, with an urban population of almost 90%, is vulnerable to a wide variety of stresses and shocks. The main climatic issues for Melbourne include heatwaves, bushfires and floods. The three levels of government - local, state and commonwealth - and hundreds of business and community organisations are part of an approach and large-scale project known as 'resilient Melbourne'.

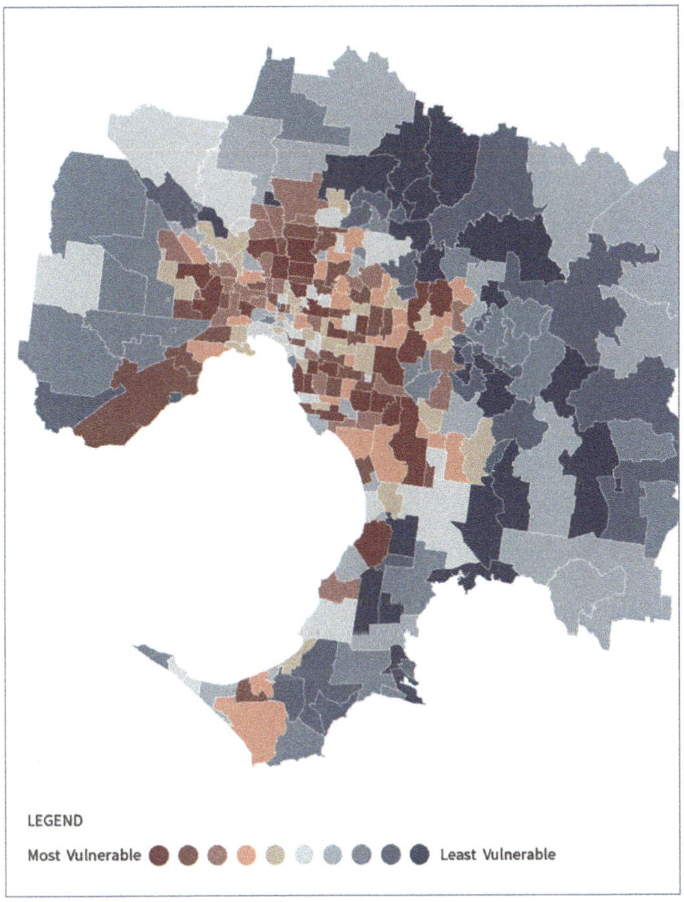

Figure 4.7: **Vulnerability of residents to extreme heat waves in Melbourne**

Source: 100 Resilient Cities, City of Melbourne Resilience Strategy. 2016. https://www.melbourne.vic.gov.au/SiteCollectionDocuments/resilient-melbourne-strategy.pdf (accessed 04/09/2021). p.52

Resilient: The capacity to recover quickly from difficulties.

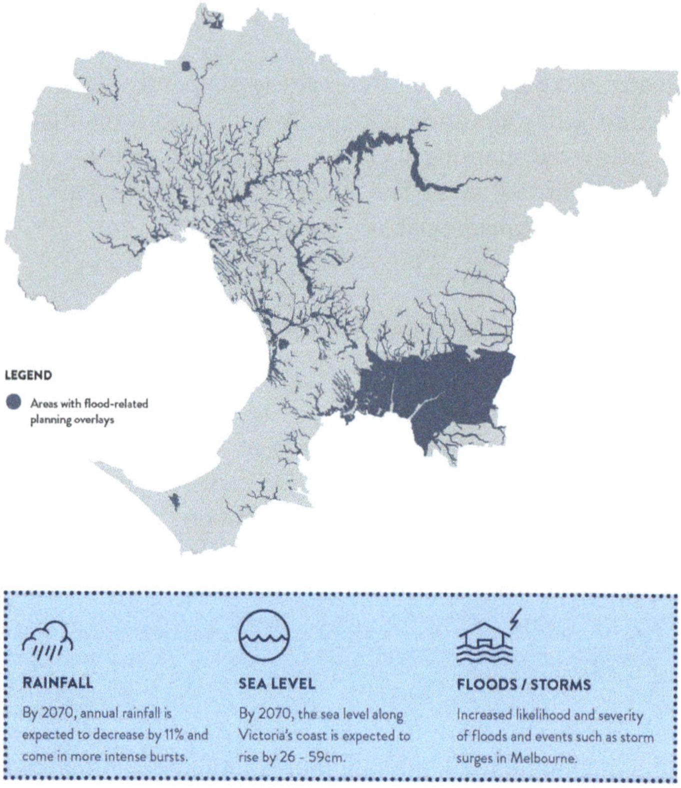

Figure 4.8: **Areas of Melbourne prone to flooding and the projected risk due to climate change**

Source: 100 Resilient Cities, City of Melbourne Resilience Strategy. 2016. https://www.melbourne.vic.gov.au/SiteCollectionDocuments/resilient-melbourne-strategy.pdf (accessed 04/09/2021). p.53

The scheme has identified four main stages - Adapt, Survive, Thrive, and Embed - with specific targets to achieve resilience in the city. The first target directly relates to mitigating the threats of climate change, as it has been identified that some of Melbourne's local government areas already have among the lowest urban tree canopy ratios in Australia. On Melbourne's fringes, new suburbs are being built on what was once arable greenfield sites. These new hard surfaces absorb the sun's heat and result in an urban heat island (see section 3.1): peak inner metropolitan temperatures can be up to 7°C higher than those in surrounding rural areas. This is all occurring in the context of climate change, which will cause more frequent heatwaves, droughts, and extreme rainfall.

The first 'adapt' plan is the metropolitan urban forest strategy. It is intended to:

- Provide shadier, cooler metropolitan areas
- Lower flood risk for people and assets
- Ensure less storm water and nutrients enter waterways and Port Phillip Bay
- Reduce CO_2 in the atmosphere
- Promote ecosystem health and biodiversity.

	ACTION	PAGE	ACTION TYPE	RESILIENCE OBJECTIVES: Stronger together	Our shared places	A dynamic economy	A healthier environment
ADAPT	Metropolitan urban forest strategy	62	Flagship Action	Related goal	Related goal		Primary resilience goal
ADAPT	Integrated Water Management Framework	68	Supporting Action		Related goal		Primary resilience goal
ADAPT	The Neighbourhood Project	74	Supporting Action	Related goal	Primary resilience goal		Related goal
ADAPT	New apartments trial for public housing residents	78	Supporting Action	Related goal	Primary resilience goal		
ADAPT	Local Government Renewables Group Purchasing	82	Supporting Action	Related goal	Primary resilience goal		Related goal
ADAPT	National Climate Resilience and Adaptation Strategy	61	Local Aligned Action		Related goal	Related goal	Primary resilience goal
ADAPT	Association of Bayside Municipalities Adaptation Planning	73	Local Aligned Action		Related goal		Primary resilience goal
ADAPT	Working with Melbourne's tertiary and further education colleges and upper secondary schools to support STEM education	87	Local Aligned Action	Related goal		Primary resilience goal	
ADAPT	Ask Izzy	81	Local Aligned Action	Primary resilience goal		Related goal	
SURVIVE	An emergency management community resilience framework for Victoria	90	Flagship Action	Primary resilience goal	Related goal	Related goal	Related goal
SURVIVE	Understanding drivers of community resilience	94	Supporting Action	Primary resilience goal	Related goal		
SURVIVE	Community-based resilience compendium	100	Supporting Action	Related goal	Related goal	Related goal	Related goal
SURVIVE	Innovative insurance	104	Supporting Action	Primary resilience goal		Related goal	
SURVIVE	IBM Smarter Cites Challenge and Twitter Data	109	Local Aligned Action	Primary resilience goal		Related goal	
SURVIVE	Flood Management Strategy – Port Phillip and Westernport	93	Local Aligned Action	Primary resilience goal			Related goal
SURVIVE	Multicultural water safety and settlement	98	Local Aligned Action	Primary resilience goal	Related goal		
SURVIVE	'Refuge': the role of art and culture in preparedness	103	Local Aligned Action	Primary resilience goal	Related goal		Related goal
SURVIVE	Building preparedness through relationships in Koori and culturally and linguistically diverse communities	107	Local Aligned Action	Primary resilience goal	Related goal		
THRIVE	The metropolitan cycling network	112	Flagship Action	Related goal	Primary resilience goal		Related goal
THRIVE	Community-led neighbourhood renewal and development pilot projects	116	Supporting Action	Related goal	Primary resilience goal		
THRIVE	Citymart Challenge – involving citizens in mobility and transport	120	Supporting Action	Related goal	Primary resilience goal	Related goal	
THRIVE	Young and Resilient Living Labs	126	Supporting Action	Primary resilience goal	Related goal	Related goal	
THRIVE	STEM Mentoring Melbourne	130	Supporting Action	Related goal		Primary resilience goal	
THRIVE	Innovative business models	134	Supporting Action	Related goal		Primary resilience goal	Related goal
THRIVE	VicHealth Mental Wellbeing Strategy	111	Local Aligned Action	Primary resilience goal	Related goal		
THRIVE	Melbourne Metro Rail Project	125	Local Aligned Action	Related goal	Primary resilience goal	Related goal	Related goal
THRIVE	Preventing Violence Together	129	Local Aligned Action	Primary resilience goal	Related goal		
THRIVE	30-year Infrastructure Strategy for Victoria	137	Local Aligned Action	Related goal	Primary resilience goal	Related goal	Related goal
EMBED	The Resilient Melbourne Delivery Office	142	Supporting Action	Related goal	Related goal	Related goal	Related goal
EMBED	City of Melbourne chair in resilient cities	148	Supporting Action	Related goal	Related goal	Related goal	Related goal
EMBED	Resilience Training for Local Government	152	Supporting Action	Related goal	Related goal	Related goal	Related goal
EMBED	Working to embed resilience and build social cohesion	151	Local Aligned Action	Primary resilience goal	Related goal	Related goal	Related goal

Flagship Action · Supporting Action · Local Aligned Action · Primary resilience goal · Related goal

Figure 4.9: **The 'Resilience Melbourne' action plan**

Source: 100 Resilient Cities, City of Melbourne Resilience Strategy. 2016. https://www.melbourne.vic.gov.au/SiteCollectionDocuments/resilient-melbourne-strategy.pdf (accessed 04/09/2021). p. 59

SECTION 4.4 PRACTICE

1. Referring to figures 4.7, 4.8 and 4.9 describe why Melbourne is attempting to become a 'resilient city'.

....

....

....

....

2. Using two 'actions' from Figure 4.9 on page 55, explain how they may mitigate the climate threats to Melbourne.

....

....

....

....

....

....

4.5 Resilient City Design: Geopolitical Risks

Resilience can also take the form of mitigating geopolitical risks, such as protection from natural hazards or managing political unrest. For example, the city of Christchurch (New Zealand) has targeted communities, buildings, and infrastructure to be better prepared to withstand catastrophic events following the large earthquake in 2010 and the subsequent aftershocks.

A regular threat to congested urban places in the twenty-first century has been terrorism, and there have been various attacks at religious centres, shopping malls, Christmas markets, and concerts worldwide. After a truck was driven into a crowd at a Christmas market in Berlin in 2016, the police chief, Klaus Kandt, pointed out that with so many potential targets – 2,500 such markets in Germany and 60 in Berlin alone – it was impossible to reduce the risk to zero.

> This is an interesting podcast by the Friends of Europe thinktank, from their 'Terror and the City' event, which discusses approaches to boosting urban resilience to violent extremism. https://soundcloud.com/friendsofeurope/terror-and-the-city (accessed 16 February 2019)

The following strategies have been proposed in major European cities:

- Invest in armed soldiers to protect potential targets (Paris and Brussels have both done this)
- Barriers to pedestrianised areas/walkways to slow down vehicles (e.g. the large concrete blocks outside the Houses of Parliament in London)
- Redesigning streets and access roads to prevent access or reduce speed of vehicles
- A complex and dynamic mix of intelligence, law enforcement, public awareness, and community relations
- Increased shared intelligence between cities and countries (the attacker at the Berlin Christmas market was captured in Milan a few days later).

Figure 4.10 shows an infographic summarising efforts by EU cities to become more resilient to terrorist threats.

Figure 4.10: **Attempts various EU cities have been making in search of geo-political resilience**

Source: Debating Europe, 'How can we make our cities terror-proof?'. 2018. www.debatingeurope.eu/2018/02/15/can-make-cities-terror-proof/#.W-Kv_ZNKg2w (accessed 04/03/2019). Courtesy of Debating Europe.

SECTION 4.5 PRACTICE

1. Outline what it means to be a resilient city.

....

....

....

2. Discuss the effectiveness of measures taken by EU cities to be more resilient to terror attacks.

....

....

....

....

....

....

4.6 Ecocity Design

The following definition of an ecocity was adopted on 20 February 2010, in Vancouver, Canada:

> *An ecocity is a human settlement modelled on the self-sustaining resilient structure and function of natural ecosystems. The ecocity provides healthy abundance to its inhabitants without consuming more (renewable) resources than it produces, without producing more waste than it can assimilate, and without being toxic to itself or neighbouring ecosystems. Its inhabitants' ecological impact reflect planetary supportive lifestyles; its social order reflects fundamental principles of fairness, justice and reasonable equity.*
>
> Ecocity Builders and the International Ecocity Framework and Standards (IEFS) advisory team. 2010.

You can measure and find out how to reduce your own footprint here: https://www.footprintcalculator.org/

One of the main aims of an ecocity is to reduce its **ecological footprint**. Currently we are running at a global deficit as the world average of around 2.8gha (global hectares) compares to an available **biocapacity** of around 1.7gha. This means we are using more

Ecological footprint: The measurement of the amount of land and water a population requires to produce the resources it consumes and to absorb its waste under prevailing technology.

Biocapacity: The capacity of the earth to generate an ongoing supply of renewable resources for a population and to absorb its waste.

resources than the planet can sustain. Cities tend to have higher ecological footprints than their rural counterparts (see Figure 4.11); therefore, urban planners have a responsibility to design more eco-friendly cities.

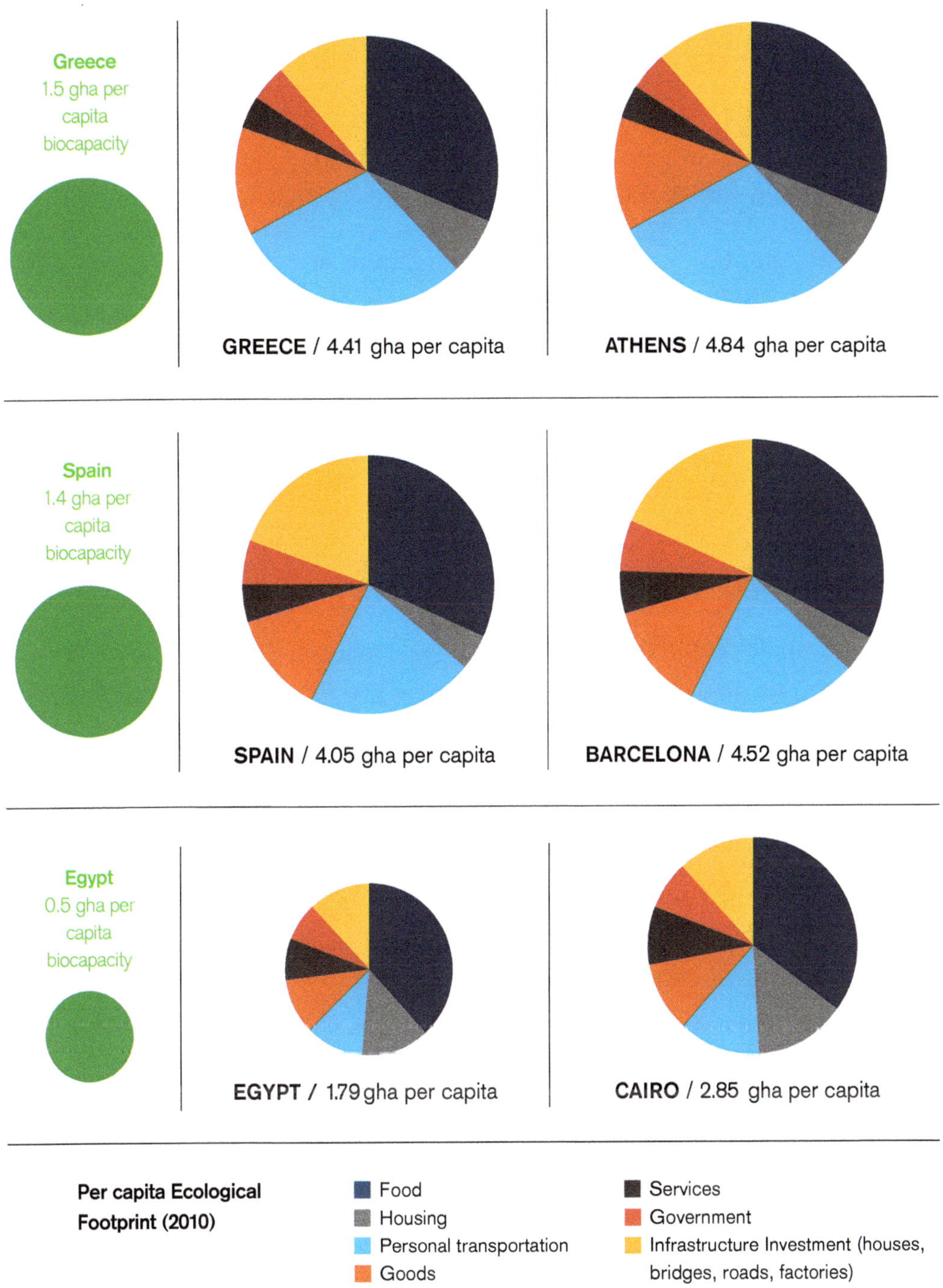

Figure 4.11: **Ecological footprints of three Mediterranean countries and a major city from each**

Source: Global Footprint Network, www.footprintnetwork.org/our-work/cities/ (accessed 14/09/2021)

CASE STUDY

Växjö, Sweden (Population: 66,275)

In 1991 this city in Southern Sweden became the first in the world to declare its intention to become fossil-fuel free. A BBC documentary in 2007 labelled Växjö the 'greenest city in Europe' and it has clearly worked hard to live up to the accolade, adopting a sustainable, **circular system**.

The city makes use of the pine forests and lakes in southern Sweden and waste wood from the forests provide 90% of its heat and a quarter of its electricity. The rest of Växjö's electricity comes from locally produced small hydro, wind, biogas, and solar plus imports from Sweden's grid, which is mostly powered by hydro and nuclear as there is heavy taxation on oil, coal and gas.

All municipal buses run on biogas from its food waste and supermarkets and canteens are full of organic and locally grown food. The city also has extensive cycle routes, pedestrianised streets and promotes electric cars (Slavin, 2015).

The result is impressive: CO_2 levels were cut by 58% in the period 1993–2016 (Trevisan, 2018).

CASE STUDY

Vancouver, Canada (Population: 631,490)

In 2011 Vancouver released its 'Greenest City 2020 Action Plan' which had an overall goal of reducing its Ecological Footprint by 33% (of 2006 levels) by 2020 and achieve one-planet living by 2050. This was as a result of findings that Vancouver residents had an ecological footprint three times larger than what was available per person worldwide. Furthermore, a new goal of zero waste/zero carbon by 2020 was set in 2015, and in 2016 the Vancouver City Council passed the Zero-Emission Building Plan. In 2017 Vancouver began a 'Neighbourhood Footprint Campaign' to help the city move closer to its goals (Global Footprint Network, 2017).

More updates on the action plan can be found here: https://vancouver.ca/files/cov/greenest-city-2020-action-plan-2015-2020.pdf

There are 10 main goals of the Action Plan.

Green Economy: double the number of green jobs and businesses with green operations

Climate Leadership: require all new buildings built after 2020 to be carbon neutral

Green Buildings: reduce CO_2 emissions in existing buildings by 20% over 2007 levels

Green Transportation: reduce driving and make the majority of trips by foot, bicycle, and public transit traffic

Zero Waste: reduce solid waste going to landfills to 50% of 2008 levels

Access to Nature: increase accessibility of green parks, greenways, and other green space, so that all Vancouver residents live within 5 minutes of these spaces

Lighter Footprint: reduce Vancouver's ecological footprint by 33% over 2006 levels

Circular system: Similar to a circular economy, an urban circular system aims to minimise inputs and outputs through processes such as recycling, as opposed to a linear system where inputs and outputs are high and products are disposed.

Clean Water: increase water quality and reduce water consumption per capita by 33% from 2006 levels

Clean Air: increase air quality, measured against Metro Vancouver and World Health Organization guidelines

Local Food: increase amount of locally grown food

Successes so far include an additional 28km of cycle paths on the Vancouver seawall, greenhouse gas emissions (GHGs) have seen an 18% decrease per capita from 2007-2015 and cigarette recycling bins have been placed around the city.

SECTION 4.6 PRACTICE

1. Discuss the concept of an 'ecocity'.

....

....

....

....

2. Compare and contrast the strategies of two cities in their attempt to reduce their ecological footprint (discounting their population size).

....

....

....

....

....

....

4.7 Smart City Design: Retrofitting Technology to Older Settlements

Smart cities are where urban places take advantage of increasing technology to make a more liveable environment. They often embrace similar ideals to eco and resilient cities and are generally considered to be sustainable. Many cities have brought in some form of technology with the aim of becoming 'smarter' from the introduction of robotic police in Dubai to the installation of CCTV in the city slums of Mumbai. Drones are a typical feature of a smart city and they have been used to manage crime and respond to the sound of gunfire in Louisville, USA.

> Try to combine your case studies, particularly for this topic. Smart city concepts are becoming increasingly common in attempts to build resilient and eco cities.

According to a recent study, the number of smart cities across the world will have quadrupled between 2013 and 2025. India, for example, is planning to spend US$1.13 billion on 100 of them (Deign, 2014).

CASE STUDY

Barcelona

Barcelona has undergone many methods to redesign the city and use technology to make it more liveable.

- The bus system has been revamped, from 108 'arbitrary routes' to 28 lines in a grid.
- Lamp posts double up as Wi-Fi access points.
- Sensors are used to monitor and manage traffic.
- Smart streetlights monitor air quality and noise.
- Smart sensors are also used to counter problems with drought.
- LED streetlights activate on movement.
- Smart garbage bins only empty when full.
- Smart parking technology show free parking spots in the city, reducing congestion.

SECTION 4.7 PRACTICE

1. Distinguish between a 'smart city' and an 'ecocity'.

....

....

....

2. Comment on the measures Barcelona has taken to become a smart city.

....

....

....

....

4.8 Smart City Design – Purpose Built Settlements

The move to design a 'smart city' from scratch is ambitious, but it may be easier to achieve than retrofitting. Well-documented examples of this include Songdo (South Korea) and Masdar City (Abu Dhabi).

This short video shows a 'drone's eye view' of Songdo: https://www.youtube.com/watch?v=eO3-zwkplRo

CASE STUDY

Belmont, Arizona

Bill Gates bought 25,000 acres of land in Arizona to build a 'Smart City' in 2017 at a cost of $80 million and a potential population size of 180,000 people. Although some sceptics see it 'as simply a potential test ground for Microsoft products' Gates sees the city, 'as a chance to build information networking into the bedrock of any future development there' (Hilburg, 2017). The proposed community will be called Belmont, located in the state's West Valley, about 45 minutes west of downtown Phoenix.

> *The intention is to create 'a forward-thinking community with a communication and infrastructure spine that embraces cutting-edge technology, designed around high-speed digital networks, data centres, new manufacturing technologies and distribution models, autonomous vehicles and autonomous logistics hubs.*
>
> Belmont Partners

It will have office, commercial, retail, residential and school zones, all built around a flexible infrastructure model. A proposed I-11 freeway would connect Belmont to Las Vegas, Nevada, opening up significant opportunity for commerce (Grant, 2017). Figure 4.12 demonstrates the aspects that a purpose built smart city may consider, taking into account SEEP factors.

This interactive page from the BBC visualises a city in 2050 – https://www.bbc.com/news/technology-23524249

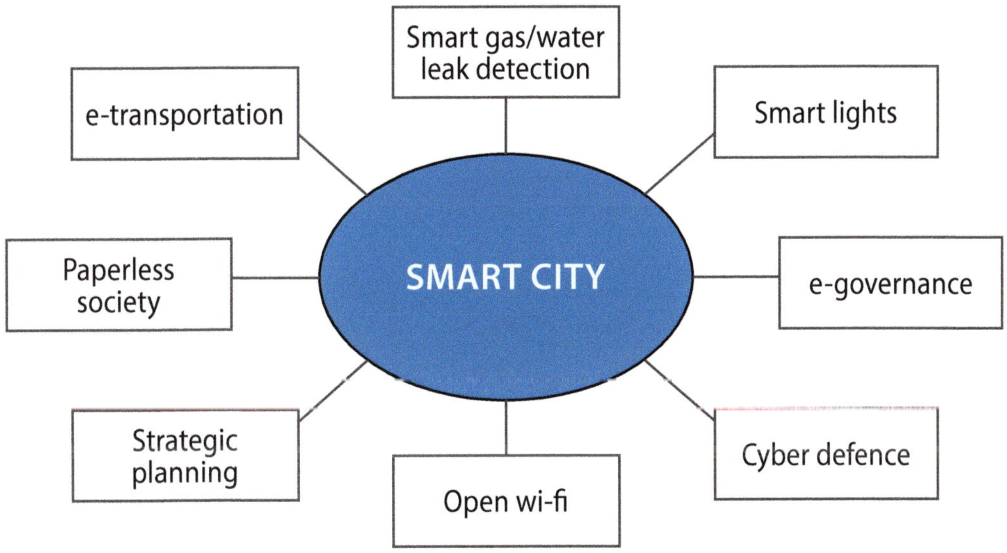

Figure 4.12: **Aspects to consider for a purpose built smart city**

SECTION 4.8 PRACTICE

1. Referring to examples, describe the benefits of a purpose built smart city.

....

....

....

....

2. 'A resilient city must also be "smart" to achieve sustainability'. To what extent do you agree with this statement?

....

....

....

....

3. Add any extra features a 'smart city' may consider to Figure 4.12.

Chapter 5: Mapping Exercise

The geography guide indicates that students should be able to locate and differentiate elements of the Earth's surface using direction, latitude, longitude, grid references, area references, scale, and political units. There may be a topographic map in this option in your final exam, or one of the other options you study; therefore, it is vital you practise these skills.

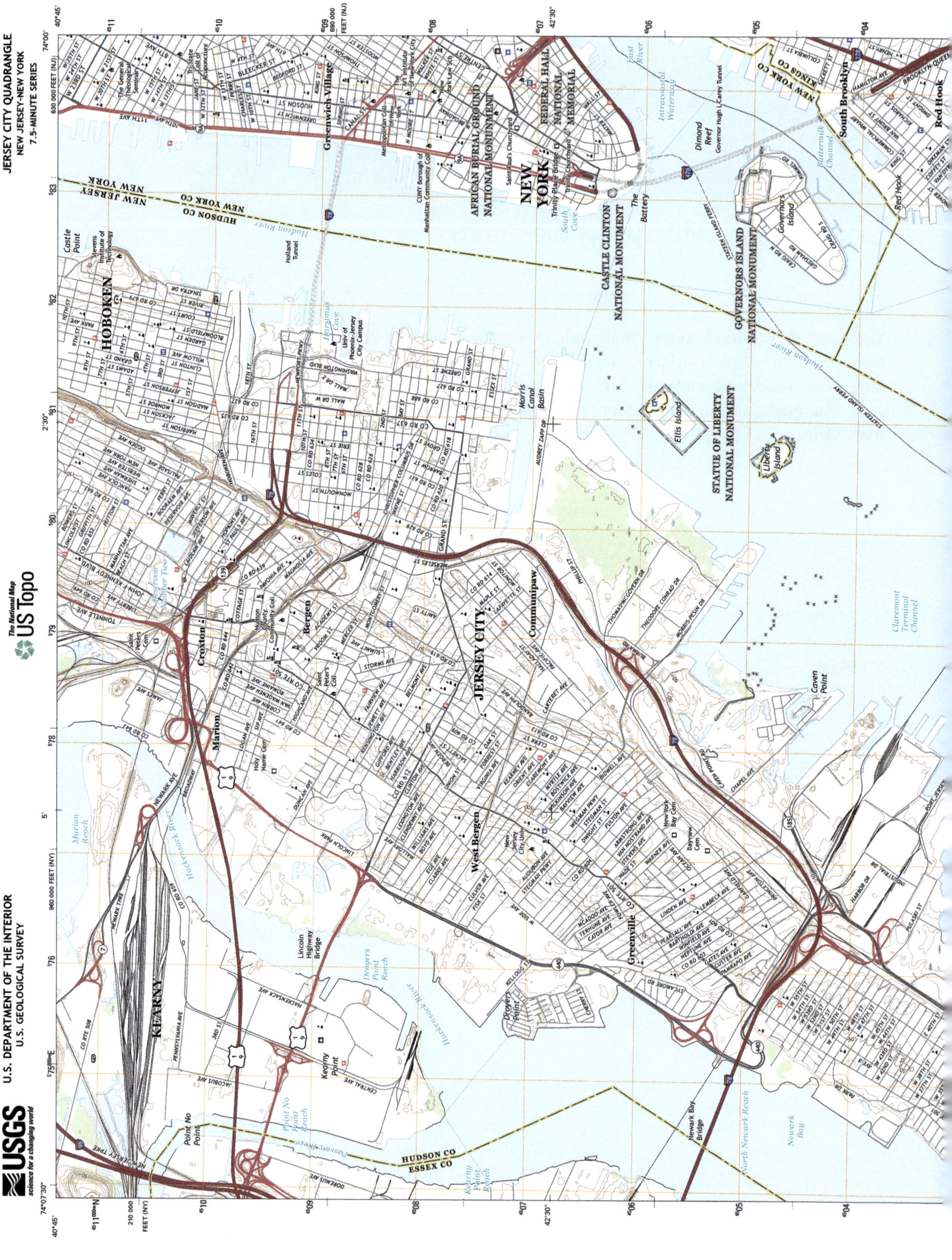
JERSEY CITY QUADRANGLE
NEW JERSEY-NEW YORK
7.5-MINUTE SERIES
U.S. DEPARTMENT OF THE INTERIOR
U.S. GEOLOGICAL SURVEY
USGS
science for a changing world
The National Map
US Topo
HOBOKEN
JERSEY CITY
NEW YORK
Greenwich Village
Castle Point
Stevens Institute of Technology
Holland Tunnel
Hudson River
NEW JERSEY
NEW YORK
HUDSON CO
NEW YORK CO
KINGS CO
AFRICAN BURIAL GROUND NATIONAL MONUMENT
FEDERAL HALL NATIONAL MEMORIAL
CASTLE CLINTON NATIONAL MONUMENT
GOVERNORS ISLAND NATIONAL MONUMENT
STATUE OF LIBERTY NATIONAL MONUMENT
The Battery
Governors Island
Ellis Island
Liberty Island
South Brooklyn
Red Hook
Buttermilk Channel
East River
Morris Canal Basin
Communipaw
Bergen
Croxton
Marion
West Bergen
Greenville
Caven Point
Claremont Terminal Channel
KEARNY
Kearny Point
Hackensack River
Passaic River
Newark Bay
Newark Bay Bridge
North Newark Reach
Point No Point
HUDSON CO
ESSEX CO
Lincoln Highway Bridge
Droyers Point
Marion Reach

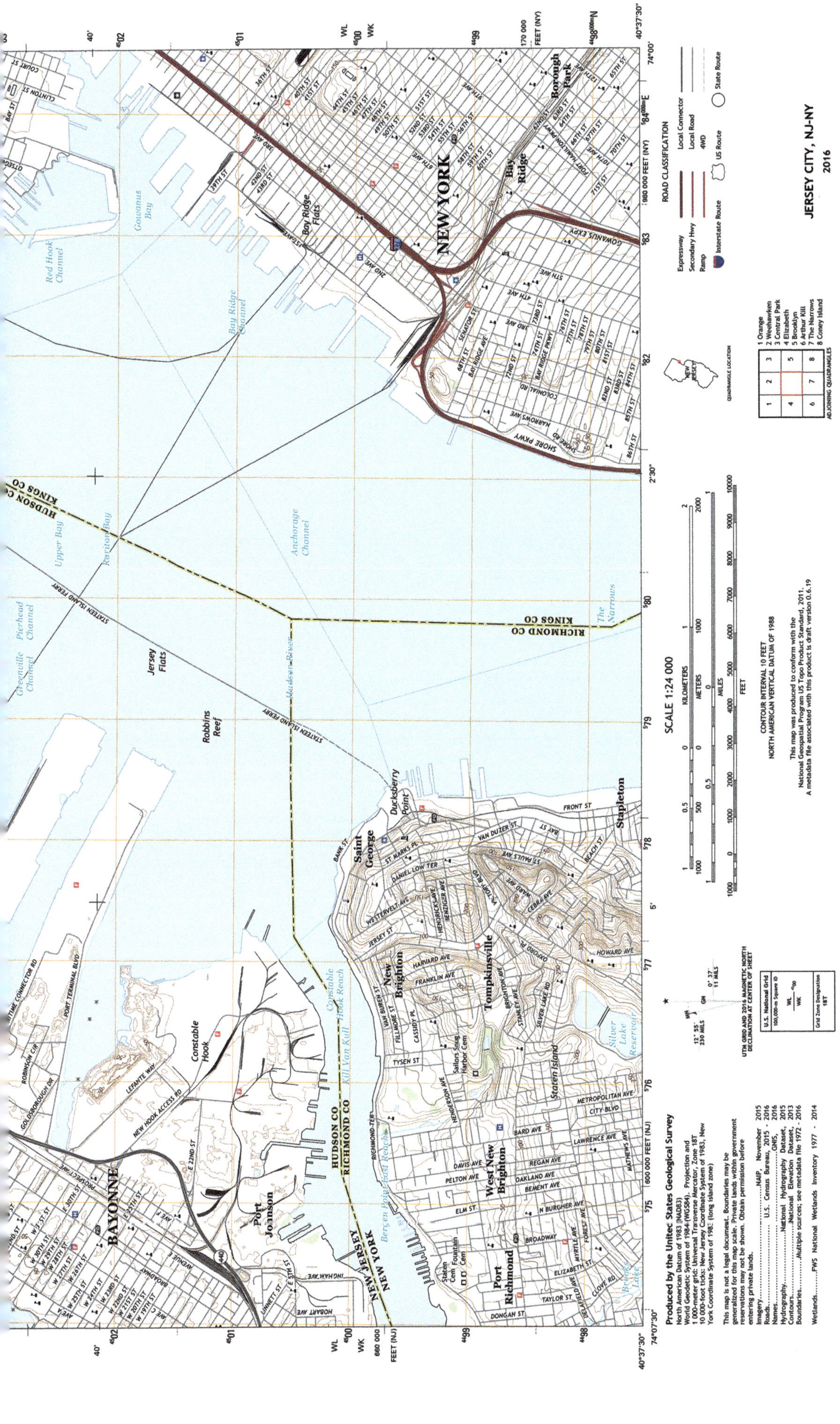

Figure 5.1: **Map of Jersey City, NJ-NY**

Source: United States Geological Survey (USGS)

MAPPING PRACTICE EXERCISE

Study the USGS extract of New Jersey/New York in Figure 5.1 on pages 66–67.

1. Estimate the distance between the Statue of Liberty (Liberty Island) and Caven Point.

2. State the direction you would be travelling from Liberty Island to reach Jersey City.

3. Identify the grid square containing New Jersey University.

4. Suggest the grid reference for St. Paul's Cathedral.

5. Estimate the height around Silver Lake Reservoir.

6. Identify an area on the map that could be considered suburban.

7. Give two reasons for your answer in question 6.

8. Taking the Staten Island Ferry from 834060 to Ducksberry Point:
   (a) Suggest the grid reference for the landing point.

   (b) State the direction of travel.

   (c) Estimate the total distance of the journey.

Chapter 6: Answers to Practice Questions

Section 1.1 practice

1. Site 3 is above the marshland and could be useful for defence. However, it is far away from water and natural resources.

   Site 4 is a route centre as it has both roads and rivers leading into the area, meaning it also has a water supply. However, being so close to water may make it more prone to flooding.

2. There are many answers for this, some examples are included below.

FUNCTION	MEANING	EXAMPLES
Industry	An urban area that developed with industry being their primary activity	Glasgow, UK Chicago, USA
Market	An urban area that developed with selling and buying goods at a market being their primary activity	Hedeby, Denmark Marktbergel, Germany
Port	An urban area that developed due to their port	Rio de Janeiro, Brazil New York City, USA
Resort	An urban area that developed due to it being a popular tourist location	Siem Riep, Cambodia Ubud, Bali, Indonesia

Section 1.2 practice

1. (a) A megacity will likely have the full variety of functions and services. Land use will also vary, as the sprawling metropolis will contain high amounts of residential areas, as well as industry and commerce.

   (b) A large town may have multiple functions but is more likely to have a primary function than a megacity. For example, the land use may be dominated by retail with commercial activity playing a secondary role. High order services are likely to exist, but not at the same scale as a megacity.

   (c) A small town is less likely to have a variety of high order services and will mainly focus on lower order services such as a supermarket and hairdresser. The primary function and land use is likely to be residential, with some open space.

2. Although all settlements generally start off as an isolated dwelling, certain factors enable some to grow faster than others. As functions and land use become more

varied, and an increasing number of people move into an area, a hierarchy or 'order of importance' can be identified. At the top may be a megacity or conurbation with the most variety of high-order services, while at the bottom there will be smaller villages with a narrow amount of low-order services.

3. The site of a city is the natural area upon which a settlement is built, and the situation is the relation of this to places and features surrounding it, such as accessibility. They both have a big influence on the function of cities. For example, a 'wet-point site' is more likely to develop functions of trade or industry, while a route-centre would be more likely to develop into a market town.
4. This model shows that most cities have a number of banking districts, industrial areas, and varying residential areas. It illustrates that a city is more spread out and generally contains more than one CBD. It also suggests there are residential suburb(s) further out from the centre and unlike the sector or concentric models, this does not necessarily follow a 'ring' pattern or specific transport routes. Therefore, with the concept of distance decay, this is probably a more accurate representation of modern day cities, where it can be difficult to determine one central place.
5.

R	esidential
I	ndustrial
C	ommercial
E	ntertainment
P	ublic buildings
O	pen space
T	ransport
S	ervices

Section 1.3 practice

1. (a) Oceania/North America

   (b) Africa
2. There are many options for this: for example, China, India, Indonesia, Bangladesh, Afghanistan, Burkina Faso, Namibia, Sudan, Somalia etc.
3. Depends on the country. For example:

   China: although having experienced huge urban growth in the twenty-first century, there are still approximately 500,000 people living a rural lifestyle. This is often an ageing population as younger citizens have moved to the cities, so this dynamic may well change in the future.

   Sudan: this country is still an LIC and therefore largely depends on agriculture as a primary economic activity. As it economically and socially develops in the future, more people may start moving to urban areas.
4. The 37 megacities are dispersed around every continent except Oceania. The highest concentration is in Asia, which has over half of them, and particularly China which appears to contain 7. Africa and North America have 3 respectively, while South America has 4. In Europe there are 4, if you count Istanbul. It is evident that many of these are coastal, although there are many anomalies here such as Moscow and Delhi. There is a cluster of megacities between the equator and the tropic of cancer, particularly in Asia.
5. Figure 1.8 shows a well-organised, planned city with tall skyscraper style buildings and perhaps smaller residential/office buildings surrounding it. There is a visible

transport infrastructure and well-maintained roofs. Some of these roofs appear to have gardens promoting urban agriculture and a more sustainable lifestyle. Figure 1.9 shows urban sprawl and evidently poorer, unfinished buildings with many open roofs. It looks quite disorganised with uncontrolled growth of vegetation and could be described as 'cluttered.' They both have high buildings which are probably used for office space as vertical zoning is common in urban environments.

6. In Egypt, they are developing a new capital (as yet unnamed) to replace the existing overcrowded Cairo. The purpose of this is to develop a cleaner environment for a new government. The city will have 5 million residents (as opposed to the 20 million in Cairo), a new parliament and palace, and low pollution levels with plenty of green spaces.
7. The rate of rural-urban migration exceeds the supply of housing that is available. This may result in unplanned, spontaneous growth of settlements as people set up slum housing in open spaces. Furthermore, the continued growth of older cities outside their limits, due to processes such as suburbanisation, has also led to this phenomenon.

Sections 1.4 and 1.5 practice

1. From this chart, one can deduce that as we move from the CBD to the rural-urban fringe, the number of high order services and offices decreases and residential, industry, and open space increases. A CBD has approximately 30% high order services and around 40% offices, compared to around 5% and 10% for the rural-urban fringe. This is not a consistent pattern though. Due to the lower cost of land at the edge of city, there are more high order services and industry than in the suburbs (almost 50% industry in the rural-urban fringe compared to around 10% in the suburbs). Furthermore, suburban locations show the highest amount of residential, attributed to both suburbanisation and the desire for families to live in more 'leafy' locations, but still accessible to the CBD. The inner city is more diverse, with a variety of residential (more likely younger people living near the centre for work) and a large amount of low order services (around 20%) to appease the residents.
2. When new migrants arrive in a city, is it natural for them to cluster together with people who speak the same language and have the same culture. Furthermore, following eras of colonisation, clusters emerged in many European cities. Another example could be the Apartheid era in South Africa, where people of black origin were forcibly kept apart from those of white origin. The discriminatory nature of this meant large, poorer clusters appeared throughout the cities in the country.
3. The theory that as you move further towards the centre of a city, the land price is higher. Therefore, only certain wealthy companies/individuals can afford the high rent of land in these places. Hence, higher order services and better accessibility are often found in the CBD.

   (a) Economic activities such as retail and industry need to be located where there is good accessibility and a suitable threshold, enabling enough people to work and use these activities. However, equally important, certainly historically, was the prevalence of water for industry and trade, flat land to build on, and an abundance of natural resources.

   (b) Residential areas tend to be in land further away from the CBD in modern HIC cities, with a better environmental quality and good access to trains/roads to get to work. In LICs, physical factors are less important as spontaneous shanty settlements can spring up almost anywhere around a city's boundaries, even on steep hillsides (such as in Rio de Janeiro). Here, human features are certainly of greater importance.

Section 1.6 practice

1. Zambia has reduced its incidence of poverty in urban areas, from around 25% in 2010 to 22% in 2015. The country has seen a marginal decrease overall due to the rise in rural areas and still stands at over 50%. Lusaka, against the theory of urban areas having less poverty, has some of the highest rates in the country.

The informal economy is the unregulated flow of money through employment not registered by the government. As education levels are low, particularly in the slum housing areas, it can be difficult for people to find official work. As the informal sector can involve construction and many 'hard labour' style jobs, it is the perfect employment for many of the people who live in the slums. Around 60% of the city are employed in this sector.

Urbanisation is still seen in many LICs as the 'avenue to development.' They are also associated with higher levels of wealth and education due to the abundance of schools, jobs and other services such as hospitals (therefore being able to treat people quicker if necessary) and hence reducing poverty levels. GDP levels are, overall, highest in those countries which are predominately urban, such as in Western Europe and rich oil states such as Qatar. Niger with under 20% of its population urban also has one of the lowest GDP per capita rates in the world at below $1,000. The flip side to this is the conditions in some of the poorest shanty settlements, which can be worse than their rural counterparts.

Section 1.7 practice

1. Lusaka and London are two very contrasting urban environments, yet both can consider poverty as a great urban stress. In Lusaka, more than 70% of the population live in slum housing centred around the CBD. They can lack access to water and basic sanitation and are frequently exposed to disease. In London there is an evident wage gap, whereby despite the overall wealth there are still high levels of poverty. Around 27% of the population are said to live in poverty in the city, mainly in the inner city area with a clear East/West divide.
2. There is a clear correlation. The 'City and East' and 'North East' regions have rates of 40%, or even more than 50% child poverty and also have the highest rates of deprivation. Anomalies can be seen, for example, in the 'South West' region where there are higher levels of child poverty compared to deprivation, where it is generally quite low.
3. There are two kinds of poverty: absolute and relative. Absolute is where a person or household has insufficient means to obtain basic services and is generally considered to be earning less than $1.90 per day. Relative poverty is where an individual or household earns less than 60% of the median income of their city/country. Poverty can also mean more than just a lack of money and directly correlates with food insecurity and a lack of access to basic services.

Section 2.1 practice

1. Delhi
2. Japan's ageing population will result in a general demographic decline in the country. Processes such as counter urbanisation may also play a role, along with the very high cost of living in Tokyo.

3. Lusaka will see the biggest percentage increase (379%), then Lagos (332%), Dhaka (324%) Delhi (315%), Shanghai (282%), and Tokyo (12.4%).
4. Push factors could be poverty or flooding. Pull factors could be better schools or the higher amount of equipped hospitals.
5. LIC cities, such as Lusaka, are currently going through the transition from a 'rural based' economy to a more industrialised society, and therefore people are drawn by the pull factors such as better paid jobs. HIC cities meanwhile have largely gone through the urban migration process, are expensive to live in, and have low levels of natural population increase.

Section 2.2 practice

1. Gentrification mainly refers to the redevelopment of housing areas and is often associated with urban renewal. It usually happens in poorer, minority communities and in areas which are suffering from urban decay. These people are then replaced by an influx of younger, wealthier groups.
2. Pros - can 'revolutionise' areas that are in serious disrepair, creating a better image for the city. It can attract new developments and services with the new population that move in.

   Cons - often gentrified areas lead to the existing residents being unable to afford the new housing prices and being 'forced out'.

   Parkdale, in Toronto, is an example of this. Zero units of affordable housing were offered in new housing developments in 2018 and the area has seen an influx of younger, wealthier residents. Locals have protested at this following new chain restaurants ('Vegandale') coming in and replacing more established, family run Indian and Tibetan restaurants.
3. There has been a suggestion that 'climate gentrification' could be on the horizon, as areas become too hot/suffer from more extreme natural disasters to live in. Wealthy people may then move to areas less subject to extremes, raising house prices. It is said this could become commonplace in southern US states such as Arizona.

Section 2.3 practice

1. While counter-urbanisation can be interpreted similarly to 'urban-rural migration' (or urban-smaller urban) by people, suburbanisation refers to the encroachment of surrounding suburban areas by an urban place and its people, physically extending the boundaries.
2. Centrifugal movements are largely caused by the desire of people to live in an area with a better environmental quality than in the CBD or inner city. The cheaper cost of land and the declining importance of distance decay are also relevant.
3. The depletion of greenfield sites and subsequent loss of biodiversity. Inner city areas can become derelict and run down, whereas zones outside of the centre may thrive. In extreme circumstances an urban area may expand so much, that surrounding cities adjoin to create a conurbation, or even a megalopolis.

Section 2.4 practice

1. Infrastructure can refer to transport, telecommunications, water supply and sanitation, and also to public buildings. For a city to develop and cater for its residents, particularly with future population projections in urban areas, these must

be of a reasonable quality and constantly upgraded and tested to ensure they meet the needs of residents.

2. In Rio de Janeiro, the transport infrastructure is very well developed, with an advanced metro system, three airports, a port and a bus, and tram network. Some of these were upgraded for the 2016 Olympic Games, but there are large disparities throughout the country. Rio is fairly advanced in terms of its television and phone; however, only 66% of residents were reported to have internet access in 2016. Sanitation varies considerably throughout the city, with reports of open sewers in favelas, compared to well-developed western style toilets in more affluent neighbourhoods.
3. Brazil has been named 'the most unequal country on earth' and a large part of this is due to the prevalence of favelas (shanty towns) in cities like Rio. These favelas are often lacking in services such as waste management and sanitation and are often said to be put to the 'bottom of priorities' for government officials. Residents also tend to suffer from lower education rates and hence a lack of formal employment and cannot afford basic services. They generally don't have any rights to their land and their homes can be bulldozed legally, as was seen in the build up to the Olympic Games in the city in 2016.

Section 2.5 practice

1. Deindustrialisation is the reduction of industrial activity or capacity in a region or economy. It is often associated with the dramatic decline in heavy industry since the 1950s in major cities in Europe and North America.
2.
   (a) Glasgow: heavily reliant on shipbuilding and heavy industries, Glasgow was unprepared for the rise in technology from the 1950s. Cars, trains and airplanes replaced sea transport and iron and steel making became mechanised. Competition and cheaper labour from emerging markets (e.g. China) also contributed to this decline.

   (b) Chicago: the Chrysler plant in Chicago mass produced airplane engines, but it was heavily impacted by the rise in globalisation and the service economy. Businesses outsourced their work to cheaper countries.
3.

	SOCIAL	ECONOMIC	DEMOGRAPHIC
Glasgow	Overcrowding in the areas: Gorbals had 85,000 people on 2% of the city Rats and street gangs due to increasing poverty Higher crime rates	Substantial job losses having a knock-on effect on the economy Poverty was evident More investment in the service sector	Population declined from its peak of 1.1 million in the mid-1900s, to around 620,000 in 2017
Chicago	Mainly associated with the economic factors High levels of unemployment and emigration were evident	27,500 job losses at the Chrysler plant in less than a year Closure of many manufacturing firms	Population dropped from a 1950 peak of over 3.2 million, to around 2.7 million in 2017

4. When investigating the consequences of deindustrialisation, it is important to look at them together; therefore, socio-economic consequences may be a better term. The economic issues of unemployment and subsequent lower wages and GDP directly

related to the increases of poverty and emigration in both Chicago and Glasgow. Therefore, it is difficult to say the social problems were 'worse' despite reports of higher crime and violence.

Section 3.1 practice

1. An urban heat island refers to a concentrated urban area that is hotter than that surrounding it. Generally this will be the CBD and cities with larger skyscrapers tend to feel the effects more, sometimes up to 5 degrees warmer. The tall buildings can act as wind tunnels, funnelling the wind in a canyon effect and reducing the breeze. Furthermore, the dark surfaces in urban areas absorb the heat, particularly at night when it is slow to escape. Heat from transport systems and 24 hour working offices can add to this. Add in the increase of population density during the day, the lack of green spaces for shade and high amounts of air pollution then the microclimate can be a considerable stress.
2. As per Figure 3.1, with annotations taken from the answer to question 1 above.

Section 3.2 practice

1. Delhi is regarded as the most polluted city in the world and air pollution has a huge effect on the people. With incredibly high levels of PM10 particular matter, about 5 times the recommended safe limit on average, air pollution in the city has led to increased rates of lung cancer, diabetes and general headaches and sore throats. It is estimated that there are 2.2 million children who have irreversible lung damage. There is also the case of 'pollution migrants' where some people (particularly elderly who are more prone to sickness) leave the city during the colder months to escape.
2. There are more cities featured (6) in Asia than any other continent. The lowest rates are found in the more economically developed cities such as London and New York. The 'highest' cities on the map are all in Asian and African cities. Every continent has at least one city representation.
3. Delhi has high urban growth rates and is an incredibly congested, heavily populated megacity. New York, meanwhile, is a much older city with lower growth rates. The city has invested in green spaces and put restrictions on polluting cars and industries. These are not in place in Delhi as India continue to go through its rapid industrialisation phase.

Section 3.3 practice

1. Traffic congestion has many causes such as old, inappropriate narrow roads in HICs, poor public transport facilities, or a lack of alternatives such as bike lanes. The variety of different types of vehicle on the road (Rickshaws in Dhaka, for example) mean traffic chaos can be evident in cities worldwide.
2. Brussels was known as the congested capital of Europe as recently as 2013 but, despite its continued issues, has moved down the rankings considerably. Greater awareness in the city promoted through car free days, the installation of extended tram lines and more frequent services, the provision of electric bikes and scooters, and the pedestrianisation of an area outside the 'Bourse' in the CBD have all helped to reduce the average reported number of hours waiting in congestion from over 80 in 2013 to 39 in 2017. Despite this, however, the city can still be gridlocked at peak times and has been recognised as having high amounts of NO_2.

Section 3.4 practice

1. Ownership rights and governmental policies are among many factors leading to contested land use. The battle between environmentalists and those wanting to continue building on greenfield sites for economic gain can cause considerable conflict. Rapid global urbanisation has meant an increase in these conflicts.

2. In Danyame neighbourhood in Kumasi, Ghana, there is a confusing ownership of land. Three different bodies, including 'chiefs', claim to have control. These chiefs have been known to sell land privately without proper permission from the government. The overall city has quintupled in size since the 1970s and major conflicts regarding the use of green space (or non-use) have intensified with more rural-urban migration in recent years. The lack of a specific 'green plan' means conflicts over this space are likely to continue. Barra de Tijuca in Rio de Janeiro was home to many slum dwellers in the early part of the twenty-first century. Following Rio's successful bid for the Olympic Games in 2016, it was reported that thousands of people were forced to relocate, sometimes for transport infrastructure, but also to help improve the city's image and build facilities for the Olympics that will likely be dormant in the future.

3. (a) The depletion of green space

   This can lead to, among others, a loss of biodiversity and inadvertently have a role in climate change. As green space disappears to be replaced with urban infrastructure, more absorption from solar insolation is likely, intensifying microclimates.

   (b) The demolition of slum housing

   This leads to the less advantaged, poorer populations being forced out of their homes. Some in Barra, Rio de Janeiro, have been known to be relocated as far as 40 kilometres away, separating families and destroying cultures.

Section 3.5 practice

1. London, compared to the rest of the UK, has a very high crime rate. Levels from 2017 indicate that the city had double the amount of knife crime offences than the next nearest region. There are also distinct boroughs/wards that see higher amounts of crime, such as Westminster and Camden.

2. There is a clear correlation between deprivation and crime in London. The lowest crime rates (in the 50s per 1000 of the population) are seen in the outer boroughs such as Harrow and Sutton, which also have lower rates of deprivation. Islington, Tower Hamlets, and Hackney appear to fit the trend of high crime and high deprivation. The highest crime rates are in Westminster, which at over 212 is almost double that of the next highest in Camden. Camden has high deprivation in the south-east, but Westminster is somewhat of an anomaly with its low deprivation. The high crime here could be attributed to this being the CBD and tourist centre with high amounts of pickpocketing and minor crimes.

3. Having a heavy police force and investing money into security and advanced technology are obvious ways in which crime levels can be reduced. One scheme that has seen some success is that employed by the Scottish government, who have treated crime as a 'disease' and have been diagnosing the problem and treating it, working together with police, teachers and social workers. Homicide rates were halved in the years 2007–2017.

Section 4.1 practice

1. The patterns of urban growth are that Asian countries will continue to see huge rates of urbanisation in the likes of India and China (together with Nigeria, they will have around 35% of the projected urban growth up to 2050), but it is in Africa, currently the least urban, where rates will increase the highest. Many new cities and megacities are expected in the continent by 2050. Trends will see rural global population stabilise and start to decline to just over 3 billion by 2050, with almost 70% of the world's population residing in urban areas (around 6.7 billion people). This is around a 5.7 billion increase in urbanites in 100 years.

2. African countries are going through the rapid urbanisation phase and are projected to continue to do so. At 57% rural in 2017, it is the least urbanised continent and are following the patterns of other continents. Major urban centres, such as Lagos in Nigeria and Lusaka in Zambia, offer higher salaried jobs and the draw of better education and health services, hence encouraging rural–urban migration.

3. Many HIC cities, such as Chicago and Glasgow, have been through the process of deindustrialisation, leading their populations to decline. Add in the processes of counter-urbanisation and suburbanisation, and the desire for more 'green' living conditions, then the centripetal forces of urban renewal and gentrification are countered, meaning growth rates generally remain low or stable.

Section 4.2 practice

1. Africa is going to urbanise the fastest, meaning there is going to be increasing pressure on urban services, such as transport and waste management, and a likely rise in slum conditions and a continuation of high poverty levels. Asia will also urbanise rapidly. Meanwhile, all other continents will have slow growth, possibly resulting in urban stresses such as urban decay and a decline in overall services such as hospitals. On the other hand, slow growth can lead to more sustainable living and less unemployment, and fast growth can (under the right circumstances) improve GDP.

2. Urban population growth rates are lower in 4 of the 7 countries in 2015 than they were in 1960. The biggest exception to this is Niger which rises from 4.54% to 5.44%, despite an almost 1% dip in 1990. The UK and India see modest rises. The most dramatic drop is seen in Poland, by over 5% to a negative growth rate. Zambia also lowers over 3%, although this has risen by over 1.5% since 1990.

3. Niger and Zambia's growth rates have fluctuated due to the technological drive of the early 2000s, with the rise in service jobs enticing rural migrants as they did in the 1960s.

   Poland has gone through rapid economic growth and the subsequent demand for more green space by inhabitants and lowering of the birth rates could explain their fall. The migration of millions following membership of the EU in 2005 could also have contributed to this.

   The UK has risen, perhaps due to processes of urban renewal and gentrification in areas of Glasgow and London.

4. Although some cities may have high populations, it depends on the proportion they have grown as to whether they can achieve a sustainable lifestyle. A good example is Tokyo - this is actually forecast to shrink in size but will still have a population in excess of 35 million people. Lusaka meanwhile will only have about 5 million

by 2030, but this is a huge proportional rise in its urban population from under 2 million in 2010.

5. (a) 4

   (b) Between 27–28 million

   (c) Between 5-6 million

6. Between 1950 and 2000, India's urban population has risen from around 65 million to around 290 million. That's over four times the amount of urbanites in 50 years. It is projected to triple from the year 2000 to 2050, from around 290 million to 880 million. (This will still only be around half of India's projected population by then.)

Section 4.3 practice

1. The city will see an overall rise in population. In 2000 this city had a very youthful population, with almost 20% of the population below the age of 4. In 2050, however, this had almost halved at just over 10%. In 2000, life expectancy was low, with around 23% of the whole population considered elderly dependents over 65. Meanwhile this figure has increased to over 40% by 2050. The highest age group is 0–4 in 2000 but is 50–54 in 2050. This indicates most of these children survived and the city has gone through a demographic dividend as they have moved into the economically active. The structures are similar in that the male/female ratio is fairly steady throughout.

2. This city has clearly been through economic development as birth rates have dropped significantly as, possibly, more women go to work and have children later. Better health care facilities have enabled people to live longer and lower rates of rural–urban migration mean the city has grown relatively slowly in 50 years.

Section 4.4 practice

1. Melbourne suffers from extreme heatwaves (in the west) and flooding (particularly in the south-east), and these are likely to intensify with climate change. It is forecast to have an 11% decrease in rainfall, but a large rise in sea level by 2070; therefore, the city must find ways to become more resilient.

2. Action 1: The 'local government renewables group purchasing' may lead to less waste produced and disposed. This in turn can mitigate the threats of climate change as less production generally means less greenhouse gases in the atmosphere, lowering the threat of heat waves.

   Action 2: 'Working with schools' will help to promote awareness and educate students and staff to work together to combat climate threats, perhaps setting up environmental committees or running 'eco-schools' projects.

Section 4.5 practice

1. A resilient city is one that can recover quickly from difficulties, and in the twenty-first century these include climatic and geo-political threats. For example, a quick recovery from natural disasters or responding to increased terror attacks.

2. Due to the spate of terror attacks in the twenty-first century, cities are having to be more resilient in both how they set up public spaces and in how they respond. While guarding every public site or tourist site is impossible, the increased use of armed soldiers on the streets is said to have helped limit further attacks in Paris and Brussels. To prevent more vehicle attacks, barriers have been erected in sites in London and around Christmas markets in Germany. Furthermore, shared intelligence between

countries has been vital in tracking down members of terrorist groups. They have had some success, and a number of attacks have been thwarted, but by no means has the continued threat been overcome.

Section 4.6 practice

1. An ecocity is one that focuses on environmental sustainability and is at the heart of all planning. It will aim to decrease its ecological footprint by promoting planetary supportive lifestyles, such as recycling, alternative energy and transport, and minimising waste and production.

2. Växjö in Sweden was one of the first 'ecocities' by declaring its intention to be fossil-fuel free by using waste wood from forests to provide heat and electricity, with the rest coming from renewable sources. Canteens and supermarkets are stocked up with local, organic food and buses run on biogas from food waste. This enabled the city to lower their CO_2 levels by 58%, hence dramatically lowering the footprint.

   Vancouver is another example of a city aiming to lower its footprint, aiming for a 33% decrease by 2020 with its 'Green Action Plan'. This includes zero waste; a 'green economy push' for jobs, clean water, and air; and a provision of local food. Therefore, the strategies have similarities, although Vancouver is encompassing more economic aspects into the plan and has more targets for the future, such as all buildings being carbon neutral after 2020.

Section 4.7 practice

1. A smart city aims to utilise the latest technology to enhance the liveability of an urban area. An eco-city ensures that ecological measures and environmental sustainability are at the heart of city planning. They can share characteristics however, as any smart city developments are likely to be undertaken with environmental consideration in mind. Likewise eco-cities would aim to utilise technology to help them in their efforts.

2. Barcelona is at the forefront of retrofitting technology to cities. It has Wi-Fi access points on its streetlights that in turn are operated by sensors, reducing energy waste. Air quality and noise are monitored through sensors and 'smart garbage bins' only empty when full. Congestion has also been reduced with smart parking technology showing free spots in the city.

Section 4.8 practice

1. A purpose built smart city can utilise the latest technologies without being restricted by older infrastructure. For example, the proposed 'Belmont' city in Arizona will have new manufacturing technologies and auto vehicles and flexible 'zones' all built around an interconnected network. With smart sensors able to detect gas and water leaks and air pollution, it is easier to build this from the beginning in many ways than updating older, slower systems.

2. To an extent, this statement could be considered true. Resilience refers to the act of recovering quickly and being prepared and cities have considerable geopolitical and climatic risks. To this end it is essential that resilient city planners utilise the latest technologies, such as drones and cyber technology, to aid in their mitigation efforts for terrorism or rising sea levels. However, more traditional methods, such as increasing police force and financial methods, can still play a vital role in achieving resilience.

Section 5 map interpretation

1. 2 km
2. North-west
3. 7707
4. 836073
5. 220–230 m
6. A few examples, but possibly the area around grid squares 7608, 7609 and 7709
7. It is further away from the built up area, where there is more space which is typical of suburban locations. There is greenery evident, which is not so typical of CBD areas. The road structure appears to be less dense.
8. (a) 784997

   (b) South-west

   (c) Approximately 10 km

Appendix

Glossary

Absolute poverty: Where household income is below a necessary level to maintain basic living standards, generally accepted as less than $1.90 per day.

Albedo: The total amount of reflection of solar radiation.

Bid-rent theory: The theory that the highest bidders can pay the highest rent. In urban places this indicates the central, most accessible area will have the highest rent.

Biocapacity: The capacity of the earth to generate an ongoing supply of renewable resources for a population and to absorb its waste.

Blue collar worker: Workers who are usually involved in manual labour (as opposed to white collar workers who are involved in administration of office work).

Canyon effect: When tall buildings flank both sides of city streets it can artificially create a canyon-like effect. It can affect temperature, light, wind, and air quality.

CBD: Central Business District. The town centre or downtown area, with a variety of services.

Centrifugal: An outward urban movement.

Centripetal: An inward urban movement.

Circular system: Similar to a circular economy, an urban circular system aims to minimise inputs and outputs through processes such as recycling, as opposed to a linear system where inputs and outputs are high and products are disposed.

Core-periphery pattern: The world is said to be divided into core regions (mainly HICs) and periphery regions (mainly LICs). Cities can also have core areas (usually the CBD) and peripheral areas (outskirts).

Counter-urbanisation: The movement of people away from urban areas outside of the city.

Deindustrialisation: The process of areas ceasing or dramatically lowering production of heavy manufacturing industry, associated with many European and North American cities from the 1960s.

Demographic dividend: This refers to a 'bulge' in the adult, economically active population and is generally considered a favourable economic position for a city to be in.

Deprivation: Being without a basic, necessary need.

Distance decay: The concept that as distance between places increases, interaction declines.

Ecological footprint: The measurement of the amount of land and water a population requires to produce the resources it consumes and to absorb its waste under prevailing technology.

Function: The primary role of a settlement.

GDP: Gross Domestic Product. The total value of goods and services produced within a country.

Gentrification: The redevelopment of existing inner city areas, usually focusing on residential and characterised by an influx of more affluent residents.

Globalisation: The increasing interconnectivity and interdependence of nations worldwide.

Greenfield site: An area of land that has not previously been built on.

Hierarchy: Placing items in order of importance/size.

High order services: Services that are bought less often, e.g. cars.

Impervious: Not allowing water to pass through.

Informal activity: Work done that is neither taxed nor monitored by the government.

Long-wave radiation: The energy radiating from the Earth as infrared radiation.

Megacity: A city with over 10 million people.

Megalopolis: A chain of adjacent metropolitan areas, which may merge into a continuous urban region.

Metropolitan area: A region consisting of a densely populated urban core and its less-populated surrounding territories, including the rural-urban fringe.

Millennium Development Goals: Superseded by the Sustainable Development Goals, these were 8 goals devised in the year 2000 by the UN to be achieved by 2015. They had various successes.

Outsourcing: Where a company hires another company to be responsible for existing activity, often in a different country.

Planned cities: Cities that are carefully planned from their inception, often on new, undeveloped land.

PLVI: Peak Land Value Intersection. The highest land value in the city.

Pull factors: Factors making people want to go to an area.

Push factors: Factors making people want to leave an area.

Relative poverty: The condition in which people lack the minimum amount of income needed in order to maintain the average standard of living in the society in which they live.

Resilient: The capacity to recover quickly from difficulties.

Rural-urban fringe: The 'edge of the city', often breaking into green, rural areas.

Rural-urban migration: The process of moving from a rural area to an urban area.

Site: The physical land on which a settlement is built.

Situation: The relationship with surrounding human features.

Sphere of Influence: A spatial region over which an urban place can draw its customers/visitors.

Spontaneous urban growth: Fast and unplanned urban growth. It is particularly evident in cities that are rapidly growing in LICs and MICs.

Stakeholder: A person or group with an interest or concern in something.

Suburbanisation: The outward growth of urban development which may engulf surrounding villages and towns into a larger urban area.

Threshold: A threshold population is the minimum number of people necessary before a particular good or service can be provided in an area.

Urban heat island: A metropolitan area that's a lot warmer than the rural areas surrounding it.

Urban renewal/regeneration: Improving and redeveloping an urban area, often brownfield inner city sites or part of the CBD.

Urban sprawl: The spread of urban growth into undeveloped land, often with little regard for urban planning.

Urbanite: A person who resides in an urban area.

Vertical zoning: Due to the high price of land, developers tend to build upwards in the CBD, perhaps with different land uses in the same skyscraper, such as retail and commerce.

Bibliography

Adjei Mensah, C. 2014. 'Destruction of Urban Green Spaces: A Problem Beyond Urbanization in Kumasi City (Ghana)'. *American Journal of Environmental Protection*. 3, 1, pp. 1–9

Alston, Philip. 2018. 'Statement on Visit to the United Kingdom'. OHCHR. www.ohchr.org/Documents/Issues/Poverty/EOM_GB_16Nov2018.pdf (accessed 24 April 2019)

Ananthanarayanan, N., and Joydeep Thakur. 2017. 'Delhi air pollution interrupts play in India vs Sri Lanka Test, visitors wear masks on field.' *Hindustan Times*. 3 December. www.hindustantimes.com/india-vs-sri-lanka-2017/delhi-pollution-holds-up-play-in-india-vs-sri-lanka-test-visitors-wear-masks/story-22TYF7FjcXJlkzro7xz58l.html (accessed 24 April 2019)

Banks, Martin. 2017. 'Brussels' Big Bet on Pedestrianisation: Where do we stand three years on?' *The Brussels Times*. 10 November. www.brusselstimes.com/business/9547/brussels-big-bet-on-pedestrianisation-where-do-we-stand-three-years-on (accessed 24 April 2019)

Barnes, Robert. 2018. 'Constructing the New Capital of Egypt'. Construction Review Online. 10 July. constructionreviewonline.com/2017/11/new-capital-egypt-construction-project (accessed 24 April 2019)

Barrington-Bush, Liam. 2018. 'Toronto's Many Faces of Gentrification'. *NowToronto*. 17 October. nowtoronto.com/news/gentrification-toronto-pakdale/ (accessed 24 April 2019)

BBC Bitsize. n.d. 'Human environments'. www.bbc.com/bitesize/topics/zcqnvcw (accessed 24 April 2019)

BBC News. 2018. 'London violent crime could take "a generation" to solve' 5 November. www.bbc.com/news/uk-england-london-46095279 (accessed 24 April 2019)

BBC World News. 'Equatorial Guinea government moves to new city in rainforest.' 8 February. www.bbc.com/news/world-africa-38911573 (accessed 24 April 2019)

Bennet, Oliver. 2018. 'Why Egypt Is Building A Brand New Mega Capital City.' *Independent*. 10 September. www.independent.co.uk/news/long_reads/egypt-capital-city-cairo-architecture-the-new-administrative-capital-a8521981.html (accessed 24 April 2019)

Bhagat, Ram. 2014. 'Urban Migration Trends, Challenges and Opportunities in India.' International Organisation for Migration. www.researchgate.net/publication/282851921_Urban_Migration_Trends_Challenges_and_Opportunities_Background_Paper_World_Migration_Report_2015_International_Organisation_for_Migration (accessed 24 April 2019)

Biggest US Cities. 2017. 'Chicago, Illinois Population History 1840–2017'. https://www.biggestuscities.com/city/chicago-illinois (accessed 24 April 2019)

Buntz, Brian. 2016. 'The World's 5 Smartest Cities.' *IoT World Today*. 18 May. www.iotworldtoday.com/2016/05/18/world-s-5-smartest-cities (accessed 24 April 2019)

Burke, Jason. 2017. 'How can cities protect their citizens against vehicle attacks?' *The Guardian*. 18 August. www.theguardian.com/world/2017/aug/18/how-cities-protect-citizens-vehicle-attacks (accessed 24 April 2019)

Carneiro, Julia. 2014. 'Favela life: Rio's city within a city.' BBC News. 9 June. www.bbc.com/news/world-latin-america-27635554 (accessed 24 April 2019)

Chhapia, Hemali. 2017. 'At slumtorials, CCTV cameras are new class monitors.' *The Times of India*. 22 March. https:/timesofindia.indiatimes.com/city/mumbai/at-slumtorials-cctv-cams-are-new-class-monitors/articleshow/57723480.cms (accessed 24 April 2019)

CIA. n.d. 'The World Factbook'. www.cia.gov/library/publications/the-world-factbook/fields/2219.html (accessed 24 April 2019)

City of Vancouver. 2012. 'Greenest City 2020 Action Plan.' https://vancouver.ca/files/cov/Greenest-city-action-plan.pdf (accessed 23/04/2019)

Costa, Ana. 2018. 'Brazil's Great Infrastructure Opportunity.' Brink News. 27 June. http://www.brinknews.com/brazils-great-infrastructure-opportunity (accessed 24 April 2019)

Cuadros, Alex. 2016. 'The Broken Promise of the Rio Olympics.' *The Atlantic.* 1 August. www.theatlantic.com/international/archive/2016/08/building-barra-rio-olympics-brazil/493697/ (accessed 24 April 2019)

Deign, Jason. 2014. 'Retrofitting Smart Cities.' Cisco. 17 September. newsroom.cisco.com/feature-content?articleId=1489176&type=webcontent (accessed 24 April 2019)

Debating Europe. 2017. 'How can we make our cities terror-proof?' 15 February. www.debatingeurope.eu/2018/02/15/can-make-cities-terror-proof/#.XA4hVGhKg2x (accessed 24 April 2019)

De la Fuente, Alejandro, *et al.* 2015. 'Mapping Subnational Poverty in Zambia.' World Bank Group. documents.worldbank.org/curated/en/766931468137977527/pdf/952760WP0Mappi0mbia0Report00PUBLIC0.pdf (accessed 4 March 2019)

Demographia. 2019. 'Demographia World Urban Areas.' http://www.demographia.com/db-worldua.pdf (accessed 24 April 2019)

Desjardins, Jeff. 2017. 'How the World's Smartest Cities are Being Built.' Visual Capitalist. 15 August. www.visualcapitalist.com/smartest-cities/ (accessed 24 April 2019)

Duttagupta, Ishani. 2018. 'Delhi's severe air pollution is frightening everyone, forcing sick to relocate.' *Indian Economic Times.* 18 November. economictimes.indiatimes.com/news/politics-and-nation/the-capital-air-that-is-frightening-everyone-forcing-sick-to-relocate/articleshow/66670422.cms (accessed 24 April 2019)

Ecocity Builders. n.d. 'What Is An Ecocity?' ecocitybuilders.org/what-is-an-ecocity/ (accessed 24 April 2019)

Elliott, Larry. 2013. 'UK shadow economy worth £150bn.' *The Guardian.* 4 June. www.theguardian.com/business/2013/jun/04/uk-shadow-economy (accessed 24 April 2019)

Expatica. 2004. 'Brussels 'choking' with traffic congestion.' 20 April. www.expatica.com/be/news/country-news/Brussels-choking-with-traffic-congestion_117821.html (accessed 24 April 2019)

Frigola, Ricard. 2018. 'The Current Urban Reality of Rio de Janeiro: Light and Dark Sides of the Olympic Legacy.' IE. 13 February. www.ie.edu/corporate-relations/insights/the-current-urban-reality-of-rio-janeiro-light-and-dark-sides-of-the-olympic-legacy/.

Global Footprint Network. n.d. 'City and Regional Work.' www.footprintnetwork.org/our-work/cities/ (accessed 24 April 2019)

Global Footprint Network. n.d. 'Has humanity's Ecological Footprint reached its peak?.' 9 April. www.footprintnetwork.org/2018/04/09/has_humanitys_ecological_footprint_reached_its_peak/ (accessed 24 April 2019)

Global Footprint Network. 2017. 'Vancouver kicks off neighborhood Footprint campaign.' 20 February. www.footprintnetwork.org/2017/02/20/vancouver-kicks-off-neighborhood-footprint-campaign/ (accessed 24 April 2019)

Grant, Kinsey. 2017. 'Bill Gates Buys 25,000 Acres of Land in Arizona to Build "Smart City".' TheStreet. 14 November. www.thestreet.com/story/14387483/1/bill-gates-buys-land-in-arizona-to-build-smart-city.html (accessed 24 April 2019)

Guardian. 2012. 'Deprivation and poverty in London: get the data.' 12 April. www.theguardian.com/uk/datablog/2012/apr/12/deprivation-poverty-london (accessed 24 April 2019)

Harris, Gardiner. 2015. 'Holding Your Breath in India.' *New York Times.* 29 May. www.nytimes.com/2015/05/31/opinion/sunday/holding-your-breath-in-india.html?_r=0 (accessed 24 April 2019)

Hilburg, Jonathan. 2017. 'Bill Gates buys 25,000 acres of Arizona desert to build a smart city.' *The Architects Newspaper.* 13 November. archpaper.com/2017/11/bill-gates-arizona-smart-city/ (accessed 24 April 2019)

Hope, Alan. 2018. 'Air quality in Brussels "alarmingly bad".' *The Brussels Times.* 31 May. www.brusselstimes.com/brussels/11486/air-quality-in-brussels-alarmingly-bad-shows-inquiry (accessed 24 April 2019)

INRIX. 2018. 'INRIX Global Traffic Scorecard.' inrix.com/scorecard/ (accessed 24 April 2019)

Kalumbi, John. 2013. 'Population and Demographic Projections 2011-2035.' Central Statistical Office. www.zamstats.gov.zm/phocadownload/Zambia%20Census%20Projection%202011%20-%202035.pdf (accessed 24 April 2019)

Krishnan, Manisha. 2018. 'Capitalist Vegans Are Tearing Apart This Toronto Neighbourhood' 7 August. www.vice.com/en_ca/article/xwk5kq/toronto-parkdale-vegans-protest-vegandale (accessed 24 April 2019)

Lartey, Jamiles. 2018. 'Nowhere for people to go: who will survive the gentrification of Atlanta?' *The*

Guardian. 23 October. www.theguardian.com/cities/2018/oct/23/nowhere-for-people-to-go-who-will-survive-the-gentrification-of-atlanta (accessed 24 April 2019)

Mackintosh, Thomas. 2018. 'Can Scottish police help stop violent deaths in London?' 24 January. www.bbc.com/news/uk-england-london-42690960 (accessed 24 April 2019)

Magsi, Habibullah. 2017. 'Land Use Conflicts in the Developing Countries: Proximate Driving Forces and Preventive Measures' *The Pakistan Development Review*. 56, 1. www.researchgate.net/publication/316513451_Land_Use_Conflicts_in_the_Developing_Countries_Proximate_Driving_Forces_and_Preventive_Measures (accessed 24 April 2019)

McCarthy, Niall. 2018. 'Delhi Has The Worst Air Quality In The World.' Statista. 11 May. www.statista.com/chart/13814/delhi-has-the-worst-air-quality-in-the-world/ (accessed 24 April 2019)

McKay, Johnston. n.d. '1950s To The Present Day.' The Glasgow Story. www.theglasgowstory.com/story/?id=TGSFO (accessed 24 April 2019)

Mfula, Chris. 2018. 'Zambia declares curfew in Lusaka slum struck by cholera.' Reuters. 7 January. www.reuters.com/article/us-zambia-cholera/zambia-declares-curfew-in-lusaka-slum-struck-by-cholera-idUSKBN1EW0MZ (accessed 24 April 2019)

Modern Ghana. 2010. 'KMA demolishes "Abinkyi slum."' 27 May. www.modernghana.com/news/277442/kma-demolishes-abinkyi-slum.html (accessed 24 April 2019)

NDTV. 2018. '"Pollution In Delhi Same As Smoking 15-20 Cigarettes A Day," Says Expert.' 3 November. www.ndtv.com/delhi-news/delhi-air-delhi-pollution-delhi-air-quality-pollution-in-delhi-same-as-smoking-15-20-cigarettes-a-da-1942435 (accessed 24 April 2019)

Neslen, Arthur. 2018. 'Air pollution is "biggest environmental health risk" in Europe.' 11 September. www.theguardian.com/environment/2018/sep/11/air-pollution-is-biggest-environmental-health-risk-in-europe (accessed 24 April 2019)

Norgard, Matthew. 2015. 'Deindustrialisation in Chicago and its effects, 1970-1990.' Chrysler Village History Project. 29 October. chryslervillageproject.wordpress.com/2015/10/29/deindustrialization-in-chicago-and-its-effects-1970-1990 (accessed 24 April 2019)

Metlink. n.d. 'Urban Heat Island Introduction.' www.metlink.org/other-weather/urban-heat-islands/urban-heat-island-background/#!prettyPhoto (accessed 24 April 2019)

Osava, Mario. 2016. 'No Medals for Sanitation at Rio Olympics.' *Inter Press Service*. 27 July. www.ipsnews.net/2016/07/no-medals-for-sanitation-at-rio-olympics/ (accessed 24 April 2019)

Perves, Bilal. 2016. 'This map shows the incredible growth of megacities' World Economic Forum. July 2016. www.weforum.org/agenda/2016/07/this-map-shows-the-incredible-growth-of-megacities (accessed 24 April 2019)

Phillips, Dom. 2016. 'Rio 2016: Favela residents being evicted days ahead of Olympics.' *The Independent*. 2 August. www.independent.co.uk/news/world/americas/rio-2016-olympics-favela-shanty-towns-evicted-brazil-a7168221.html (accessed 24 April 2019)

Population Pyramid. n.d. 'Urban population growth (annual %).' www.populationpyramid.net/hnp/urban-population-growth/2015/ (accessed 24 April 2019)

Presser, Lizzie. 2016. 'What's behind the huge fall in deprivation in east London? And no, it's not gentrification.' *The Guardian*. 12 January. www.theguardian.com/society/2016/jan/12/what-behind-deprivation-east-london-newham-unemployment (accessed 24 April 2019)

Railway Technology. n.d. www.railway-technology.com/projects/rio-light-rail-transit-system-rio-de-janeiro/ (accessed 24 April 2019)

RioDeJaneiro.com. n.d. 'Rio de Janeiro History.' www.riodejaneiro.com/v/history/ (accessed 24 April 2019)

Rosaia, Jessie. 2015. 'Investigating an Asian Mega City Delhi.' http://megacitydelhitask.weebly.com/ (accessed 24 April 2019)

Slavin, Terry. 2015. 'What can the world learn from Växjö, Europe's self-styled greenest city?' *The Guardian*. 25 November. www.theguardian.com/cities/2015/nov/25/what-can-the-world-learn-from-vaxjo-europes-self-styled-greenest-city (accessed 24 April 2019)

SmartCitiesWorld news team. 2017. 'Smart cities services worth $225bn by 2026.' SmartCitiesWorld. 26 April. www.smartcitiesworld.net/news/news/smart-cities-services-worth-225bn-by-2026-1618 (accessed 24 April 2019)

Southern African Regional Poverty Network. n.d. 'Zambia: five worst deprived districts on the basis of selected poverty-related indicators.' sarpn.org/CountryPovertyPapers/Zambia/Strategy/povertyrelatedindicators.pdf (accessed 24 April 2019)

Statista. n.d. 'Weekly reach of selected media platforms in Greater Rio de Janeiro, Brazil in 2016.' www.statista.com/statistics/696513/media-reach-rio-de-janeiro-brazil/ (accessed 24 April 2019)

Streatfield, Peter Kim. 2008. 'Korail Slum in Dhaka' [image] in 'Population Challenges for Bangladesh in the Coming Decades.' *Journal of Health Population and Nutrition*. www.researchgate.net/figure/Korail-slum-in-Dhaka_fig4_23295854 (accessed 24 April 2019)

Trevisan, Emma. 2018. 'Växjö, the greenest city in Europe.' Ecobnb. 18 August. ecobnb.com/blog/2018/08/vaxjo/ (accessed 24 April 2019)

Trust for London. 2017. 'London's Poverty Profile 2017.' www.trustforlondon.org.uk/publications/londons-poverty-profile-2017 (accessed 7 November 2018)

United Nations, Department of Economic and Social Affairs, Statistics Division. n.d. 'Democratic Yearbook System.' unstats.un.org/unsd/demographic-social/products/dyb/index.cshtml (accessed 24 April 2019)

United Nations. n.d. 'Sustainable Development Goals.' www.un.org/sustainabledevelopment/ (accessed 24 April 2019)

United Nations, Department of Economic and Social Affairs, Population Division. 2015. World Urbanization Prospects: The 2014 Revision. https://esa.un.org/unpd/wup/publications/files/wup2014-report.pdf (accessed 24 April 2019)

United Nations, Department of Economic and Social Affairs, Population Division. 2018. World Urbanization Prospects: The 2018 Revision. population.un.org/wup/Publications/Files/WUP2018-KeyFacts.pdf (accessed 24 April 2019)

Vermeersch, Laurent. 2014. 'Five reasons Belgium has the worst traffic in Europe.' *The Guardian.* 28 August. www.theguardian.com/cities/2014/aug/28/belgium-worst-traffic-europe-brussels-antwerp-congestion (accessed 24 April 2019)

Vidal, John. 2018. 'Why is Rio de Janeiro finding it so hard to clear up its waste?' *The Guardian.* 3 August. www.theguardian.com/global-development/2016/aug/03/why-is-rio-de-janeiro-finding-it-so-hard-to-clear-up-its-waste-olympic-games (accessed 24 April 2019)

Vincent, Christophe. 2018. 'Stifling effect of nitrogen oxide on Belgium visible from space' *The Brussels Times*, 29 Oct. 2018, www.brusselstimes.com/belgium/12964/the-stifling-effect-of-nitrogen-oxide-on-belgium-visible-from-space.

Wakefield, Jane. "Tomorrow's Cities: Dubai and China roll out urban robots." BBC, 10 June 2018, www.bbc.com/news/technology-41268996.

Washwatch. n.d. 'Brazil Statistics.' washwatch.org/en/countries/brazil/summary/statistics (accessed 24 April 2019)

Williams, Colin. 2004. 'Small Businesses in the Informal Economy: Making the Transition to the Formal Economy.' Small Business Council. www.academia.edu/200717/Small_Businesses_in_the_informal_Economy (accessed 24 April 2019)

Woetzel, Jonathan, *et al.* 2018. 'Urban world: Meeting the demographic challenge in cities.' *McKinsey and Company.* www.mckinsey.com/featured-insights/urbanization/urban-world-meeting-the-demographic-challenge-in-cities (accessed 24 April 2019)

World Bank. 2019. 'World Bank Country and Lending Groups.' datahelpdesk.worldbank.org/knowledgebase/articles/906519 (accessed 24 April 2019)

World Population Review. 2019. 'Delhi Population 2019.' worldpopulationreview.com/world-cities/delhi-population (accessed 24 April 2019)

World Population Review. 2019. 'World City Populations 2019.' worldpopulationreview.com/world-cities/ (accessed 24 April 2019)

Yaw Oduro, Charles, *et al.* 2014. 'Analyzing Growth Patterns of Greater Kumasi Metropolitan Area Using GIS and Multiple Regression Techniques.' *Journal of Sustainable Development.* 7, 5. http://dx.doi.org/10.5539/jsd.v7n5p13 (accessed 24 April 2019)

Central Statistical Office. 2016. '2015 Living Conditions Monitoring Survey Report.' https://www.zamstats.gov.zm/phocadownload/Living_Conditions/2015%20Living%20Conditions%20Monitoring%20Survey%20Report.pdf (accessed 24 April 2019)

Resilient Cities. www.100resilientcities.org/ (accessed 24 April 2019)